# Fundamentals of Predictive Analytics with JMP®

Ron Klimberg and B. D. McCullough

support.sas.com/bookstore

The correct bibliographic citation for this manual is as follows: Klimberg, Ron, and B. D. McCullough. 2013. *Fundamentals of Predictive Analytics with JMP*®. Cary, NC: SAS Institute Inc.

**Fundamentals of Predictive Analytics with JMP**®

ISBN 978-1-61290-623-2 (electronic book)
ISBN 978-1-61290-425-2

SAS Institute Inc., SAS Campus Drive, Cary, North Carolina 27513-2414

1st printing, April 2013

SAS provides a complete selection of books and electronic products to help customers use SAS® software to its fullest potential. For more information about our e-books, e-learning products, CDs, and hard-copy books, visit **support.sas.com/bookstore** or call 1-800-727-3228.

To our families for all their support and love:

Helene, Bryan, and Steven;
and Renee, Otis, Nancy, Mimi, Paige, Thomas, and Ross.

iv

# Contents

# About This Book

## Purpose

As we are completing this book, it is early 2013 and the United States Presidential election is over, Barack Obama won and now everyone knows. We have heard about the big buzz around topics like business intelligence, business analytics, predictive analytics/data mining/data science, and big data. We heard reports like the McKinsey Global Institute's 2011 Big Data[1] report that stated by 2018 the United States alone will face a shortage of 140,000 to 190,000 people with deep analytical skills as well as 1.5 million managers and analysts to analyze big data. We are now finding out that a major reason President Obama won was because his staff analyzed data on their customers, the electorate, which gave him a competitive advantage and helped him win the election. So, now you know—using and leveraging your data cannot only provide you with a competitive advantage in your domain; it can also help win you an election.

The aim of this book is to expose you the reader, who is an educated consumer of statistics, to a set of multivariate and other modeling techniques, so that you not only become skilled at applying these techniques, but also skilled at developing the crucial skill of telling the statistical story behind the data. (The techniques include multiple regression, ANOVA, logistic regression, principal component analysis, cluster analysis, decision trees and neural networks.) This is what the 1.5 million managers who are mentioned in the McKinsey Global Institute's Big Data report will need to be able to do.

A major portion of academic programs in business analytics, statistics, and data mining (whether in the Arts and Sciences or Business Schools) begin their curriculum with a basic introduction to statistics. That course is usually followed by a data mining/predictive analytics[2] course. The basic statistics course teaches the student how to summarize data and perform basic statistical tests using data sets with perhaps one or two variables. In the basic statistics course, a statistical study is viewed as linear. That is, the student applies a technique in response to a problem statement; then the student gives the answer and is done. The subsequent data mining/predictive analytics class requires students to analyze data sets with several hundred variables.

Multivariate/real-world statistical studies are not linear; they are iterative. You try a technique, review and reflect on the results, and determine the next step. And you develop the statistical story behind the data. With such a large conceptual jump, most students are lost; they lack a fundamental understanding of statistical analysis. For example, most introductory statistics books do not address logistic regression. But a predictive analytics or data mining course usually assumes prior knowledge of logistic regression. In this textbook, we provide the bridge to take the reader from univariate/bivariate statistics to real-world multivariate statistical analysis.

Further, nearly all statistics books fail to address the critical skill of developing the statistical story. At the end of each chapter, after developing some statistical technique, the student will know what the problems are about and which technique to use—the topic covered in the chapter. We, the authors and instructors, neglect this most critical skill of knowing when or when not to use a technique. Yet, technique problems are essential to learning the mechanics of a given statistical technique. In this text, we provide these types of problems at the end of each chapter. We additionally provide several small and large data sets at the end of the book. We strongly suggest that the instructor use these data sets with one or more chapters. And, yes, instructors, please assign another data set after a chapter when it is not appropriate to use that chapter's technique with the data set that we have provided. As you repeatedly assign the data sets, a statistical story of the data develops. Additionally, we provide several large data sets that are appropriate for semester-long projects that use one or more of the book's techniques.

## Is This Book for You?

This text is written for students at the undergraduate or graduate level. Most textbooks written on statistical techniques for multivariate and data mining/predictive analytics are written at a level that is so mathematical or so non-technical that the reader remains unable to apply the technique. We believe that our book is at the level right in between—enough mathematics so that you understand what is going on, how and when to apply the technique, and how to interpret the output. This book is not a software manual; we do not cover every option for every method. Rather, we introduce the reader to the basic concepts necessary to understand each method.

Our goal is to make the reader an educated consumer who can develop the statistical story of a multivariate data set by learning the techniques of multivariate and data mining/predictive analytics and by developing the skill to understand the statistical story.

## Prerequisites

The text assumes that the reader has taken a basic introductory statistics course. One chapter reviews the fundamental concepts that you should understand from an introductory statistics course.

## Software Used to Develop the Book's Content

The primary software application used in this book is JMP statistical software, in particular JMP 10 and JMP 10 Pro. The book offers new and enhanced resources in JMP 10, including an add-in to Microsoft Excel, Graph Builder, and data mining/predictive analytics modeling capabilities. To provide a good foundation, some of the early examples use Microsoft Excel.

## Scope of This Book

The book starts with a review of basic statistics and expands on some of these concepts to include multivariate techniques. Several multivariate techniques are discussed (principal components, cluster analysis, ANOVA, multiple regression, and logistic regression). In introducing each technique, we provide a basic statistical foundation so that the reader understands when to use the technique and how to evaluate and interpret the results. Additionally, step-by-step directions are provided to guide you through an analysis using the technique. Similarly, the last few chapters of the book introduce a few more automated predictive modeling/data mining techniques (decision trees and neural networks) and concepts.

## Example Code and Data

You can access the example code and data for this book by linking to its author pages at http://support.sas.com/publishing/authors. Select the name of the author. Then, look for the cover thumbnail of this book, and select Example Code and Data to display the data that are included in this book.

For an alphabetical list of all books for which example code and data is available, see http://support.sas.com/bookcode. Select a title to display the book's example data.

If you are unable to access the code through the website, send e-mail to saspress@sas.com.

## Additional Resources

SAS offers you a rich variety of resources to help build your SAS skills and explore and apply the full power of SAS software. Whether you are in a professional or academic setting, we have learning products that can help you maximize your investment in SAS.

| Bookstore | http://support.sas.com/bookstore/ |
| Training | http://support.sas.com/training/ |
| Certification | http://support.sas.com/certify/ |
| SAS Global Academic Program | http://support.sas.com/learn/ap/ |
| SAS OnDemand | http://support.sas.com/learn/ondemand/ |

Or

| Knowledge Base | http://support.sas.com/resources/ |
| Support | http://support.sas.com/techsup/ |
| Training and Bookstore | http://support.sas.com/learn/ |
| Community | http://support.sas.com/community/ |

## Keep in Touch

We look forward to hearing from you. We invite questions, comments, and concerns. If you want to contact us about a specific book, please include the book title in your correspondence.

## To Contact the Author through SAS Press

By e-mail: saspress@sas.com

Via the Web: http://support.sas.com/author_feedback

## SAS Books

For a complete list of books available through SAS, visit http://support.sas.com/bookstore.

Phone: 1-800-727-3228

Fax: 1-919-677-8166

E-mail: sasbook@sas.com

## SAS Book Report

Receive up-to-date information about all new SAS publications via e-mail by subscribing to the SAS Book Report monthly eNewsletter. Visit http://support.sas.com/sbr.

---

[1] Throughout the text, we will use the terms data mining, predictive analytics, and predictive modeling interchangeably. In Chapters 1 and 11, we briefly discuss their differences.

[2] Manyika, J., M. Chui, B. Brown, J. Bughin, R. Dobbs, C. Roxburgh, and A. H. Byers. 2011. Big data: The next frontier for innovation, competition, and productivity, McKinsey Global Institute.

# About These Authors

 Ron Klimberg is a professor at the Haub School of Business at Saint Joseph's University in Philadelphia, PA. Before joining the faculty in 1997, he was a professor at Boston University, an operations research analyst for the Food an Drug Administration, and a consultant. His primary research interests lie in th areas of multiple criteria decision making, DEA, facility location, data visualization, and data mining. Ron was the 2007 recipient of the Tengelmanr Award for his excellence in scholarship, teaching, and research. He received I PhD from The Johns Hopkins University.

 B. D. McCullough is a professor at the LeBow College of Business at Drexel University in Philadelphia, PA. Prior to joining Drexel, he was a senior economist at the Federal Communications Commission and an assistant professor at Fordham University. His research fields include applied econometrics and time series, accuracy of statistical and econometrics software, replicability of research, and data mining. He received his PhD from the University of Texas at Austin.

Learn more about these authors by visiting their author pages, where you can download free chapters, access example code and data, read the latest reviews, get updates, and more:

http://support.sas.com/klimberg

http://support.sas.com/mccullough

# Acknowledgments

We would like to thank Shelley Sessoms and Stephenie Joyner of SAS Press for providing support from start to the finish. We would also like to thank Julie Platt, Mary Beth Steinbach, Kathy Underwood, Candy Farrell, Jennifer Dilley, Aimee Rodriquez, and Cindy Puryear of SAS Publishing for their assistance.

In reviewing the manuscript, we want to thank Paul Marovich of SAS and Russell Lavery for their detailed reviews, which substantially improved the final product. Also, we would like to thank our students, who provided significant edits and insights.

We would like to recognize and thank those individuals who provided one or more data sets:

W. O. Dale Amburgey, Saint Joseph's University
Tom Bohannon, SAS Institute
Jerry Oglesby, SAS Institute
Chuck Pirrello, SAS Institute
Keith Pride, MDRxResearch
John Stanton, Saint Joseph's University

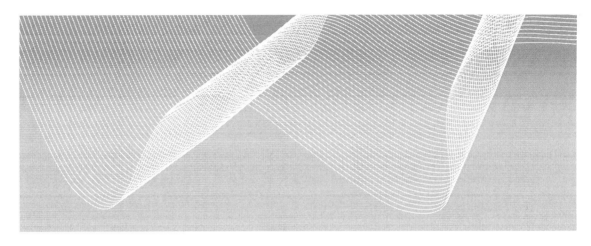

# Chapter 1

## Introduction

In 1981, Bill Gates made his infamous statement that "640KB ought to be enough for anybody."

Looking back even further, about 10 to 15 years prior to the Bill Gates statement, we were in the middle of the Vietnam era. State-of-the-art computer technology for both commercial and scientific areas at that time was the mainframe computer. A typical mainframe computer weighed tons, took an entire floor of a building, had to be air-conditioned, and cost about $3 million. Mainframe memory was approximately 512KB with disk space of about 352 MB and speed up to 1MIPS.

In 2011, only forty years later, an iPhone 4 with 32GB memory has about 9300% more memory than the mainframe and fits in our hands. A laptop with the Intel Core i7 processor has speeds up to 147,600 MIPS, about 150,000 times faster than the old mainframe, and weighs about 4 pounds. Further, an iPhone or a laptop costs significantly less than $3 million. As Ray Kurzweil, an author, inventor, and futurist has stated (Lomas, 2008):

> "The computer in your cell phone today is a million times cheaper and a thousand times more powerful and about a hundred thousand times smaller (than the one computer at MIT in 1965) and so that's a billion-fold increase in capability per dollar or per euro that we've actually seen in the last 40 years.".

Technology has certainly changed!

# Two Questions Organizations Need to Ask

## Return on Investment (ROI)

With this new and ever improving technology, most organizations (and even small organizations) are collecting an enormous amount of data. Each department has one or more computer systems. A number of organizations are now integrating these department-level systems with organization systems, such as an ERP (enterprise resource planning) system. Newer systems are being deployed that store all these historical enterprise data in what is called a data warehouse. The IT budget for most organizations is a significant percentage of the organization's overall budget and is growing. The question is:

> With the huge investment in collecting this data, do organizations get a decent ROI (return on investment)?

The answer: mixed. No matter if the size of the organization is large or small, only a limited number of organizations (yet growing in number) are using their data extensively. Meanwhile, most organizations are drowning in their data and struggling to gain some knowledge from it.

## Culture Change

How would managers respond to this question?:

What are your organization's two most important assets?

Most managers would answer with their employees and the product or service that the organization provides (they may alternate which is first or second).

The follow-up question is more challenging:

Given the first two most important assets of most organizations, what is the third most important asset of most organizations?

The answer is: the organization's data! To most managers, regardless of the size of their organizations, this would be a surprise. However, consider the vast amount of knowledge that's contained in customer or internal data. For many organizations, realizing and accepting that their data is the third most important asset would require a significant culture change.

Rushing to the rescue in many organizations is the development of business intelligence (BI) programs and groups. What is business intelligence (BI)? It seems to vary greatly depending on your background.

# Business Intelligence

Business Intelligence is considered by most people as providing IT systems, such as dashboards and online analytical processing (OLAP) reports, to improve business decision-making. Howard Dresner, considered to be the "father" of business intelligence, expanded this definition of BI in 2006 (WhatIs.com, 2006) to be a "broad category of applications and technologies for gathering, storing, analyzing, and providing access to data to help enterprise users make better business decisions. BI applications include the activities of decision support systems, query and reporting, online analytical processing (OLAP), statistical analysis, forecasting, and data mining."

Figure 1.1 presents a framework that shows the relationships of the three discipline (Information Technology, Statistics and Operations Research) components in this expanded definition of BI (Klimberg and Miori, 2010). "With this new vision, we may now characterize BI from each of three viewpoints as: business information intelligence (BII), business statistical intelligence (BSI), and business modeling intelligence (BMI). Each of the viewpoints has particular business aspects, and academically speaking, courses that are independent of the other viewpoints."

**Figure 1.1 A Framework of Business Intelligence [3]**

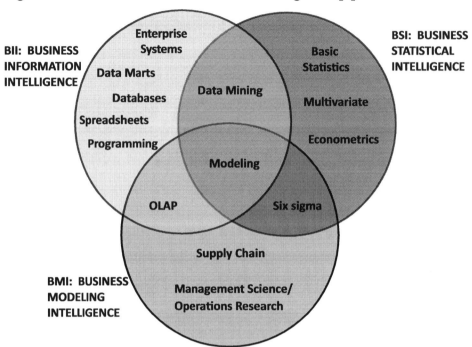

This expanded scope of BI and its growing applications have revitalized an old term: business analytics (BA). BA, in our framework, is a combination of business statistics intelligence (BSI) and business modeling intelligence (BMI): BA = BSI + BMI. Davenport (Davenport and Harris, 2007) views BA as "the extensive use of data, statistical and quantitative analysis, explanatory and predictive models, and fact-based management to drive decisions and actions."

Davenport further elaborates that organizations should develop an analytics competency as a "distinctive business capability" that would provide the organization with a competitive advantage.

The buzzword in this area of BA for about the last 25 years has been the term data mining. Until recently. The current buzzwords are predictive analytics and predictive modeling. What is the difference in these terms? As we discussed, with the many and evolving definitions of business intelligence, these terms seem to have many different yet quite similar definitions. In this text, we will not distinguish between data mining, predictive analytics, and predictive modeling and will use them interchangeably to mean or imply the same thing. In Chapter 11, we will briefly discuss their different definitions.

# Clarification

The labels business intelligence, business information intelligence, business statistics intelligence, and business modeling intelligence are not limited to the business sector. Even though these labels have "business" in front of them, the application of the techniques and tools within each of these disciplines can be and are used in the public and social sectors. In general, wherever data is collected, these tools and techniques can be applied.

# Book Focus

This book focuses on the business statistics intelligence (BSI) component of BA. In particular, given that you have likely taken a basic introduction to statistics course, this book will discuss visualization tools and multivariate statistical tools (such as, clustering analysis, principal components, and logistic regression) and processes to perform a statistical study that may include data mining/predictive analytics techniques. Some real-world examples of using these techniques are target marketing, customer relation management, market basket analysis, cross-selling, market segmentation, customer retention, improved underwriting, quality control, competitive analysis, fraud detection, and management and churn analysis. Specific applications can be found at http://www.jmp.com/software/success/. The bottom line, as reported by the KDNuggets poll (2008): the median ROI for data mining projects is in the 125–150% range.

The primary objective of this book is to provide a bridge from your introduction to statistics course to practical statistical analysis. For the most part, your introductory statistics course has not completely prepared you to move on to real-world statistical analysis.

# Introductory Statistics Courses

Most introductory statistics courses (outside the mathematics department) cover the topics of descriptive statistics, probability, probability distributions (discrete and continuous), sampling distribution of the mean, confidence intervals, one-sample hypothesis testing, and perhaps two-sample hypothesis testing, simple linear regression, multiple linear regression, and ANOVA. (Yes, multiple linear regression and ANOVA are multivariate techniques, but the complexity of the multivariate nature is for the most part not addressed in the introduction to statistics course. One main reason—not enough time!) Nearly all the topics, problems, and examples in the course are directed toward univariate (one variable) or bivariate (two variables) analysis. Univariate analysis includes techniques to summarize the variable and make statistical inferences from the data to a population parameter. Bivariate analysis examines the relationship between two variables (e.g., the relationship between age and weight).

A typical student's understanding of the components of a statistical study is shown in Figure 1.2. If the data is not available, a survey is performed or the data is purchased. Once the data is obtained, the statistical analyses are done all at once—start Excel and/or a statistical package, draw the appropriate graphs and tables, perform all the necessary statistical tests, and write up/present the results. And then you are done. With such a perspective, many students simply look at this statistics course as another math course and may not realize the importance and consequences of the material.

**Figure 1.2  A Student's View of a Statistical Study from a Basic Statistics Course**

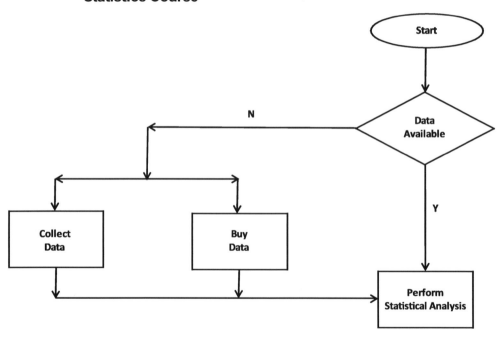

These first statistics courses provide a good foundation to introducing statistics. However, these courses provide a rather weak foundation for performing practical statistical studies. First, most real-world data are "dirty." Dirty data are erroneous data. A simple example of dirty data is if a data field/variable to represent gender is supposed to be coded as either M or F, and we find in the field the letter N or even perhaps it is blank. Learning to identify dirty data and determining corrective action is a fundamental skill to have in analyzing real-world data.

Second, most practical statistical studies have data sets that include more than two variables, called multivariate data. Multivariate analysis uses some of the same techniques and tools used in univariate and bivariate analysis as covered in the introductory statistics courses, but in an expanded and much more complex manner. Additionally, when performing multivariate analysis, you would be exploring the relationships among several variables. There are several multivariate statistical techniques and tools to consider that are not covered in a basic applied statistics course.

Before jumping into multivariate techniques and tools, students need to learn the univariate and bivariate techniques and tools that are taught in the basic first statistics course. However, in some programs this basic introductory statistics class may be the last data analysis course required or offered. Many other programs that do offer/require a second statistics course offer just a continuation of the first course and may or may not cover ANOVA and/or multiple linear regression. (Although these techniques are multivariate. we are looking for a statistics course beyond multiple linear regression.) In either case, the students are ill-prepared to apply statistics tools to real-world multivariate data. Perhaps, with some minor adjustments, real-world statistical analysis can be introduced into these programs.

On the other hand, with the growing interest in BI, BA, and data mining/predictive analytics, more programs are offering and sometimes even requiring a subsequent statistics course, in data mining/predictive analytics. So, most students jump from univariate/bivariate statistical analysis to statistical data mining/predictive analytics techniques, which include numerous variables and records. These statistical data mining/predictive analytics techniques require the student to understand the fundamental principles of multivariate statistical analysis and more so, to understand the process of a statistical study. In this situation, many students are lost, and this simply reinforces the students' view that the course is just another math course.

Even with these multivariate shortcomings, there is still a more significant multivariate concern to address. That is the idea that most students view statistical analysis as a straightforward exercise in which you sit down in front of your computer and just perform the necessary statistical techniques and tools, as in Figure 1.2. How boring! With such a viewpoint, this would be like telling someone that reading a book can simply be done by reading the book cover. The practical statistical study process of uncovering the story behind the data is exciting.

# Practical Statistical Study

The prologue to a practical statistical study is determining the proper data needed, obtaining the data, and if necessary cleaning the data (the dotted area in Figure 1.3). Answering the questions "Who is it for?" and "How will it be used?" will identify the suitable variables required and the appropriate level of detail. Depending upon who will use the results and how they will use them dictates which variables are necessary and the level of granularity. If there is enough time and the essential data is not available, then the data may be obtained either by a survey or purchased. Lastly, once the data is available, if necessary, the data should be cleaned; i.e., eliminate erroneous data as much as possible.

**Figure 1.3  The Flow of a Real-World Statistical Study**

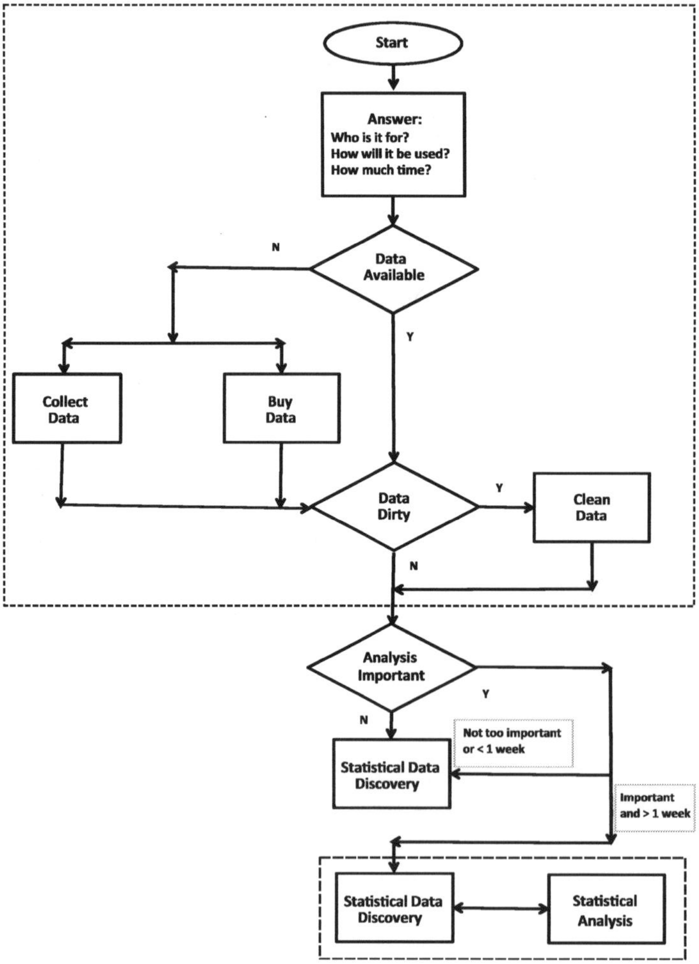

The statistical study (the enclosed dashed-line area in Figure 1.3) should be like reading a book—the data should tell a story! Part of the story is data discovery—discovery of significant and insignificant relationships among the variables and the observations in the data set. The story develops further as many different statistical techniques and tools are tried. Some will be helpful, some will not. With each iteration of applying the statistical techniques and tools, the story develops and is substantially further advanced when you relate the statistical results to the actual problem situation. As a result, your understanding of the problem and how it relates to the organization is improved. By doing the statistical analysis, you will make better decisions (most of the time). Furthermore, these decisions will be more informed so that you will be more confident in your decision. Finally, uncovering and telling this statistical story is fun!

# Plan, Perform, Analyze, Reflect (PPAR) Cycle

The development of the statistical story follows a process that we call the PPAR (PLAN, PERFORM, ANALYZE, REFLECT) cycle, as shown in Figure 1.4. The PPAR cycle is an iterative progression. The first step is to plan which statistical techniques or tools are to be applied. You are combining your statistical knowledge and your understanding of the business problem being addressed. You are asking pointed, directed questions to answer the business question by identifying a particular statistical tool or technique to use.

The second step is to perform the statistical analysis, using statistical software such as JMP. The third step is to analyze the results using appropriate statistical tests and other relevant criteria to evaluate the results. The fourth step is to reflect on the statistical results. Ask questions, like what do the statistical results mean in terms of the problem situation? What insights have I gained? Can we draw any conclusions? Sometimes, the results are extremely useful, sometimes meaningless, and sometimes in the middle—a potential significant relationship.

Then, it is back to the first step to plan what to do next. Each progressive iteration provides a little more to the story/understanding of the problem situation. This cycle continues until you feel you have exhausted all possible statistical techniques or tools (visualization, univariate, bivariate, and multivariate statistical techniques) to apply or you have enough results such that the story is completed.

**Figure 1.4 The PPAR Cycle**

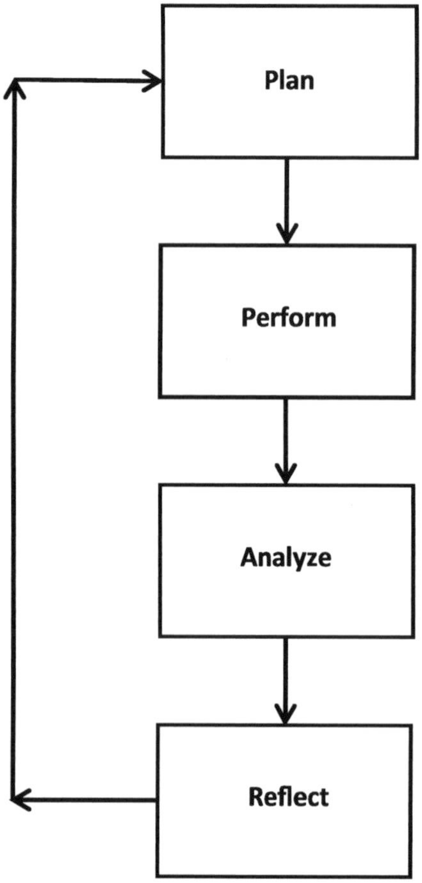

The software used in many initial statistics courses is Microsoft Excel, which is easily accessible and provides some basic statistical capabilities. However, as you advanced through the course, because of Excel's statistical limitations, you may have also used some non-professional, textbook-specific statistical software or perhaps some professional statistical software. Excel is not a professional statistics software application; it is a spreadsheet.

The statistical software application used in this book is the SAS JMP statistical software application. JMP has the advanced statistical techniques and the associated, professionally proven, high-quality algorithms of the topics/techniques covered in this book. Nonetheless, some of the early examples in the textbook use Excel. The main reasons for using Excel are twofold: (1) To give you a good foundation before we move on to more advanced statistical topics, and (2) JMP can be easily accessed through Excel as an Excel Add-In, which is an approach many will take.

In this book, we first review basic statistics and expand on some of these concepts to include multivariate techniques. Subsequently, we examine several multivariate techniques (principal components, cluster analysis, ANOVA, multiple regression, and logistic regression). The framework we will use in this book of statistical and visual methods is shown in Figure 1.5. In introducing each technique, we will provide a basic statistical foundation to help you understand when to use the technique and how to evaluate and interpret the results. Additionally, step-by-step directions are provided to guide you through an analysis using the technique. The last section of the book introduces the data mining/predictive analytics process and one of its more popular techniques (decision trees). In discussing these data mining/predictive analytics techniques, the same approach used with the multivariate techniques is taken— understand when to use it, evaluate and interpret the results, and follow step-by-step instructions. The overall objectives of the book are not only to introduce you to multivariate techniques and data mining/predictive analytics, but, to provide a bridge from univariate statistics to practical statistical analysis by instilling the PPAR cycle.

**Figure 1.5  A Framework for Multivariate Analysis**

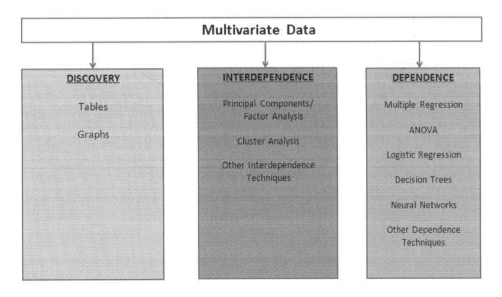

# References

Davenport, Thomas H., and Jeanne G. Harris. (2007). *Competing on Analytics: The New Science of Winning*. Cambridge, MA: Harvard Business School Press.

Klimberg, Ronald. K., and Virginia Miori. (October 2010). "Back in Business." *ORMS Today*, Vol. 37, No. 5, 22–27.

Lomas, Natasha. (November 19, 2008). "Q&A: Kurzweil on tech as a double-edged sword." *CNET News*. CBS Interactive. http://news.cnet.com/8301-11386_3-10102273-76.html.

Poll: Data Mining ROI. http://www.kdnuggets.com/polls/2008/roi-data-mining.htm.

WhatIs.com. http://whatis.techtarget.com/.

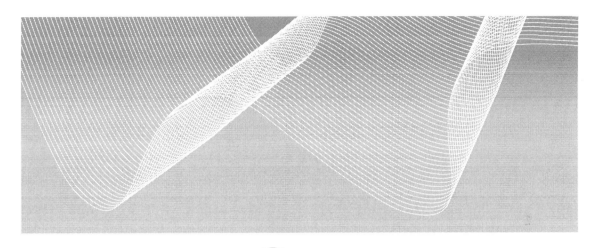

# Chapter **2**

## Statistics Review

Regardless of the academic field of study—business, psychology, or sociology—the first applied statistics course introduces the statistical foundation topics of descriptive statistics, probability, probability distributions (discrete and continuous), sampling distribution of the mean, confidence intervals, one-sample hypothesis testing, and perhaps two-sample hypothesis testing, simple linear regression, multiple linear regression, and ANOVA. Not considering the mechanics/processes of performing these statistical techniques, what fundamental concepts should you remember? We believe there are six fundamental concepts:

1.  Always take a random and representative sample

2.  Statistics is not an exact science.

3.  Understand a Z score.

4. Understand the central limit theorem (not every distribution has to be bell-shaped).

5. Understand one-sample hypothesis testing and p-values.

6. Many approaches/techniques are correct, and a few are wrong.

Let's examine each concept further.

## Always Take a Random and Representative Sample

**Fundamental Concepts 1 and 2: Always take a random and representative sample, and statistics is not an exact science.**

What is a random and representative sample (we will call it a 2R sample)? By representative, we mean representative of the population of interest. A good example to understand what we mean is state election polling. You do not want to sample everyone in the state. First, an individual must be old enough and registered to vote. You cannot vote if you are not registered. Next, not everyone who is registered votes. So, if the individual is registered, does he/she plan to vote? Again, we are not interested in the individual if he/she does not plan to vote. We do not care about their voting preferences because they will not have an impact on the election. Thus, the population of interest is those individuals who are registered to vote and plan to vote. From this representative population of registered voters who plan to vote, we want to choose a random sample. By random, we mean that each individual has an equal chance of being selected. So you could imagine that there is a huge container with balls that represent each individual identified as being registered and planning to vote. From this container, we choose so many balls (without replacing the ball). In such a case, each individual has an equal chance of being drawn.

We want the sample to be a 2R sample; but, why is that so important?

For two related reasons. First, if the sample is a 2R sample, then the sample distribution of observations will follow a similar pattern/shape as the population. Suppose that the population distribution of interest is the weights of sumo wrestlers and horse jockeys (sort of a ridiculous distribution of interest, but that should help you remember why it is important). What does the shape of the population distribution of weights of sumo wrestlers and jockeys look like? Probably somewhat like the distribution in Figure 2.1. That is, it's bimodal or two-humped.

**Figure 2.1  Population Distribution of the Weights of Sumo Wrestlers and  Jockeys**

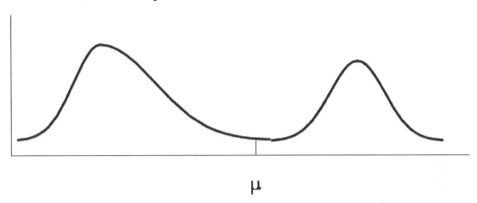

$\mu$

If we take a 2R sample, the distribution of sampled weights would look somewhat like the population distribution in Figure 2.2, where the solid line is the population distribution and the dashed line is the sample distribution.

**Figure 2.2  Population and a Sample Distribution of the Weights of Sumo Wrestlers and Jockeys**

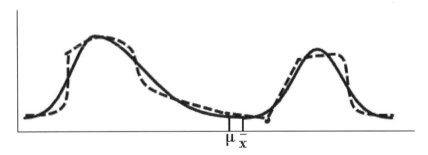

$\mu \; \bar{x}$

Why not exactly the same? Because it is a sample, not the entire population. It may differ, but just slightly. If the sample was of the entire population, then it would look exactly the same. Again, so what? Why is this so important?

## Statistics Is Not an Exact Science

The population parameters (such as the population mean, $\mu$, the population variance, $\sigma^2$, or the population standard deviation, $\sigma$) are the true values of the population. These are the values that we are interested in knowing. In most situations, the only way that we would know these values exactly was if we were to sample the entire population (or census) of interest. In most real-world situations, this would be a prohibitively large number (costing too much and taking too much time); as a result, we take a 2R sample.

Because the sample is a 2R sample, the sample distribution of observations is very similar to the population distribution of observations. Therefore, the sample statistics, calculated from the sample, are good estimates of their corresponding population parameters. That is, statistically they will be relatively close to their population parameters because we took a 2R sample.

The sample statistics (such as the sample mean, sample variance, and sample standard deviation) are estimates of their corresponding population parameters. It is highly unlikely that they will equal their corresponding population parameter. It is more likely that they will be slightly below or slightly above the actual population parameter, as shown in Figure 2.2.

Further, if another 2R sample is taken, most likely the sample statistics from the second sample will be different from the first sample; they will be slightly less or more than the actual population parameter.

For example, let's say that a company's union is on the verge of striking. We take a 2R sample of 2000 union workers. Let us assume that this sample size is statistically large. Out of the 2000, 1040 of them say they are going to strike. First, 1040 out of 2000 is 52%, which is greater than 50%. Can we therefore conclude that they will go on strike? Given that 52% is an estimate of the percentage of the total number of union workers who are willing to strike, we know that another 2R sample will provide another percentage. But another sample could produce a percentage perhaps higher and perhaps lower, and perhaps even less than 50%. By using statistical techniques, we can test the likelihood of the population parameter being greater than 50%. (We can construct a confidence interval, and if the lower confidence level is greater than 50%, we can be highly confident that the true population proportion is greater than 50%. Or we can conduct a hypothesis test to measure the likelihood that the proportion is greater than 50%.)

Bottom line: When we take a 2R sample, our sample statistics will be good (statistically relatively close, i.e., not too far away) estimates of their corresponding population parameters. And we must realize that these sample statistics are estimates, in that, if other 2R samples are taken, they will produce different estimates.

# Understand a Z Score

**Fundamental Concept 3: Understand a Z score.**

Let's say that you are sitting in on a marketing meeting. The marketing manager is presenting the past performance of one product over the past several years. Some of the statistical information that the manager provides is the average monthly sales and standard deviation. (More than likely, the manager would not present the standard deviation, but, a quick conservatively large estimate of the standard deviation is the (Max – Min)/4; the manager most likely would give the minimum and maximum values.)

Let's say the average monthly sales are $500 million, and the standard deviation is $10 million. The marketing manager starts to present a new advertising campaign which he/she claims would increase sales to $570 million per month. Let's say that the new advertising looks promising. What is the likelihood of this happening? If we calculate the Z score:

$$Z = \frac{x - \mu}{s} = \frac{570 - 500}{10} = 7$$

The Z score (and the t score) is not just a number. The Z score is how many standard deviations away that a value, like the 570, is from the mean of 500. The Z score can provide us some guidance, regardless of the shape of the distribution. A Z score greater than (absolute value) 3 is considered an outlier and highly unlikely. In our example, if the new marketing campaign is as effective as suggested, the likelihood of increasing monthly sales by 7 standard deviations is extremely low.

On the other hand, what if we calculated the standard deviation, and it was $50 million? The Z score now is 1.4 standard deviations. As you might expect, this can occur. Depending on how much you like the new advertising campaign, you would believe it could occur. So the number $570 million can be far away, or it could be close to the mean of $500 million. It depends upon the spread of the data, which is measured by the standard deviation.

In general, the Z score is like a traffic light. If it is greater than the absolute value of 3 (denoted |3|), the light is red; this is an extreme value. If the Z score is between |1.96| and |3|, the light is yellow; this value is borderline. If the Z score is less than |1.96|, the light is green, and the value is just considered random variation. (The cutpoints of 3 and 1.96 may vary slightly depending on the situation.)

## Understand the Central Limit Theorem

**Fundamental Concept 4: Understand the central limit theorem (not every distribution has to be bell-shaped).**

This concept is where most students become lost in their first statistics class; they complete their statistics course thinking every distribution is normal or bell-shaped. No, that is not true. However, if the assumptions are not violated and the central limit theorem holds, then, something called the sampling distribution of the sample means will be bell-shaped. And this sampling distribution is used for inferential statistics; i.e., it is applied in constructing confidence intervals and performing hypothesis tests. Let us explain.

If we take a 2R sample, the histogram of the sample distribution of observations will be close to the histogram of the population distribution of observations (Fundamental Concept 1). We also know that the sample mean from sample to sample will vary (Fundamental Concept 2). Let's say we actually know the value of the population mean. If we took several samples, there would be approximately an equal number of sample means slightly less than the population mean and slightly more than the population mean. There will also be some sample means further away, above and below the population mean.

Now, let's say we took every combination of sample size n (and let n be any number greater than 30), and we calculated the sample mean for each sample. Given all these sample means, we then produce a frequency distribution and corresponding histogram of sample means. We call this distribution the sampling distribution of sample means. A good number of sample means will be slightly less and more, and fewer farther away (above and below), with equal chance of being greater than or less than the population mean. If you try to visualize this, the histogram of all these sample means would be bell-shaped, as in Figure 2.3. This should make intuitive sense.

**Figure 2.3 Population Distribution and Sample Distribution of Observations and Sampling Distribution of the Means for the Weights of Sumo Wrestlers and Jockeys**

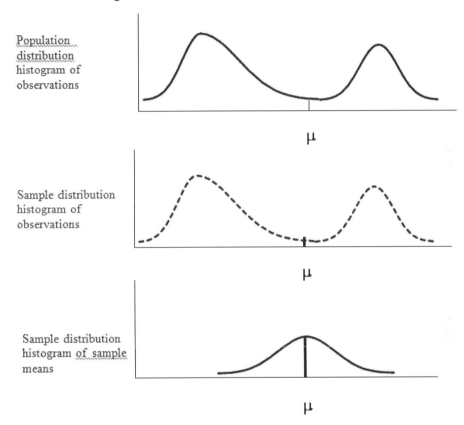

Population distribution histogram of observations

Sample distribution histogram of observations

Sample distribution histogram of sample means

Nevertheless, there is one major problem. To get this histogram of sample means, we said that every combination of sample size n needs to be collected and analyzed. That, in most cases, is an enormous number of samples and would be prohibitive. Additionally, in the real world, we only take one 2R sample.

This is where the central limit theorem (CLT) comes to our rescue. The CLT will hold regardless of the shape of the population distribution histogram of observations—whether it is normal, bimodal (like the sumo wrestlers and jockeys), whatever shape, as long as a 2R sample is taken and the sample size is greater than 30. Then, the sampling distribution of sample means will be approximately normal, with a mean of $\bar{x}$ and a standard deviation of $s/\sqrt{n}$ (which is called the standard error).

What does this mean? We do not have to take an enormous number of samples. We need to take only one 2R sample with a sample size greater than 30. In most situations this will not be a problem. (If it is an issue, you should use nonparametric statistical techniques.) If we have a 2R sample greater than 30, we can approximate the sampling distribution of sample means by using the sample's $\bar{x}$ and standard error, $s/\sqrt{n}$. If we collect a 2R sample greater than 30, the CLT holds. As a result, we can use inferential statistics; i.e., we can construct confidence intervals and perform hypothesis tests. The fact that we can approximate the sample distribution of the sample means by taking only one 2R sample greater than 30 is rather remarkable and is why the CLT theorem is known as the "cornerstone of statistics."

The implications of the CLT are highly significant. Let's illustrate the outcomes of the CLT further with an empirical example. The example that we will use is the population of the weights of sumo wrestlers and jockeys.

Open the Excel file called SumowrestlersJockeysnew.xls and go to the first worksheet called **data**. In column A, we generated our population of 5000 sumo wrestlers and jockeys weights with 30% of them being sumo wrestlers.

First, you need the Excel' Data Analysis add-in. (If you have loaded it already you can jump to the next paragraph). To upload the Data Analysis add-in:

1. Click **File** from the list of options at the top of window.

2. A box of options will appear. On the left side toward the bottom, click **Options**.

3. A dialog box will appear with a list of options on the left. Click **Add-Ins**.

4. The right side of this dialog box will now lists Add-Ins. Toward the bottom of the dialog box there will appear

   Click **Go**.

5. A new dialog box will appear listing the Add-Ins available with a check box on the left. Click the check boxes for Analysis ToolPak and Analysis ToolPak—VBA. Then click **OK**.

If you click Data on the list of options at the top of window, all the way toward the right of the list of tools will be Data Analysis.

Now, let's generate the population distribution of weights:

1. Click Data on the list of options at the top of the window. Then click Data Analysis.

2. A new dialog box will appear with an alphabetically ordered list of Analysis tools. Click Histogram and **OK**.

3. In the Histogram dialog box, for the Input Range, enter $A$2:$A$5001; for the Bin Range, enter $H$2:$H$37; for the Output range, enter $K$1. Then click the options Cumulative Percentage and Chart Output and click **OK**, as in Figure 2.4 below.

**Figure 2.4  Excel Data Analysis Tool Histogram Dialog Box**

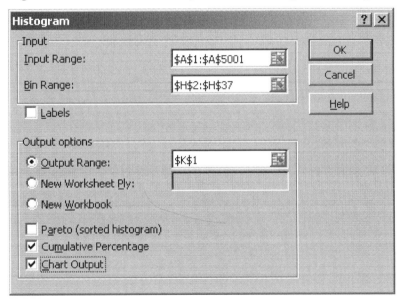

## Figure 2.5 Output of the Histogram Data Analysis Tool

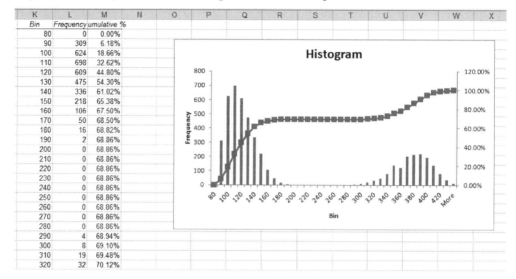

| Bin | Frequency | umulative % |
|---|---|---|
| 80 | 0 | 0.00% |
| 90 | 309 | 6.18% |
| 100 | 624 | 18.66% |
| 110 | 698 | 32.62% |
| 120 | 609 | 44.80% |
| 130 | 475 | 54.30% |
| 140 | 336 | 61.02% |
| 150 | 218 | 65.38% |
| 160 | 106 | 67.50% |
| 170 | 50 | 68.50% |
| 180 | 16 | 68.82% |
| 190 | 2 | 68.86% |
| 200 | 0 | 68.86% |
| 210 | 0 | 68.86% |
| 220 | 0 | 68.86% |
| 230 | 0 | 68.86% |
| 240 | 0 | 68.86% |
| 250 | 0 | 68.86% |
| 260 | 0 | 68.86% |
| 270 | 0 | 68.86% |
| 280 | 0 | 68.86% |
| 290 | 4 | 68.94% |
| 300 | 8 | 69.10% |
| 310 | 19 | 69.48% |
| 320 | 32 | 70.12% |

A frequency distribution and histogram similar to Figure 2.5 will be generated.

Given the population distribution of sumo wrestlers and jockeys, we will generate a random sample of 30 and a corresponding dynamic frequency distribution and histogram (you will understand the term dynamic shortly).

1. Select the 1 random sample worksheet. In columns C and D, you will find percentages that are based upon the cumulative percentages in column M of the worksheet data. Additionally, in column E, you will find the average (or midpoint) of that particular range.

2. In cell K2, enter =rand(). Copy and paste K2 into cells K3 to K31.

3. In cell L2, enter =VLOOKUP(K2,$C$2:$E$37,3). Copy and paste L2 into cells L3 to L31.

   You have now generated a random sample of 30. If you press **F9**, the random sample will change.

4. To produce the corresponding frequency distribution (and BE CAREFUL!), highlight the cells P2 to P37. In cell P2, enter the following:

   =frequency(L2:L31,O2:O37)

   THEN BEFORE pressing **ENTER**, *simultaneously* hold down and press **Ctrl, Shift,** and **Enter**.

The frequency function finds the frequency for each bin, O2:O37, for the cells L2:L31. Also, when you simultaneously hold down the keys, an array is created. Again, as you press the **F9** key, the random sample and corresponding frequency distribution changes. (Hence, this is why we call it a dynamic frequency distribution.)

5. To produce the corresponding dynamic histogram, highlight the cells P2 to P37. Click Insert from the top list of options. Click the Chart type **Column** icon. An icon menu of column graphs is displayed. Click under the left icon under the 2-D Columns. A histogram of your frequency distribution is produced, similar to Figure 2.6.

To add the axis labels, under the group of Chart Tools at the top of the screen (remember to click on the graph), click Layout. A menu of options appears below. Select **Axis Titles**→**Primary Horizontal Axis Title**→**Title Below Axis**. Type Weights and press **Enter**. For the vertical axis, select **Axis Titles**→**Primary Vertical Axis Title**→**Vertical title** and type Frequency.

**Figure 2.6  Histogram of a Random Sample of 30 Sumo Wrestler and Jockeys Weights**

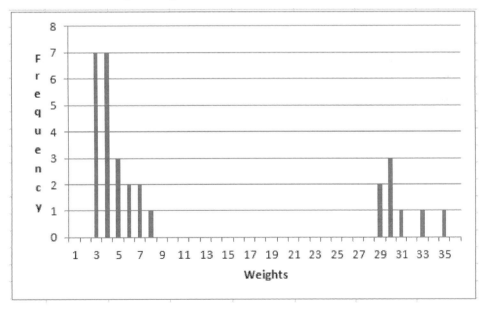

If you press **F9**, the random sample changes, the frequency distribution changes, and the histogram changes. As you can see, the histogram is definitely not bell-shaped and does look somewhat like the population distribution in Figure 2.5.

Now, go to the sampling distribution worksheet. Similarly to how we generated a random sample in the random sample worksheet, we have already generated 50 random samples, each of size 30, in columns L to BI. Below each random sample, the average of that sample is calculated in row 33. Further in column BL is the dynamic frequency distribution, and there is a corresponding histogram of the 50 sample means. If you press **F9**, the 50 random samples, averages, frequency distribution, and histogram change. The histogram of the sampling distribution of sample means (which is based on only 50 samples—not on every combination) is not bimodal, but is for the most part bell-shaped.

# Understand One-Sample Hypothesis Testing and p-Values

**Fundamental Concept 5: Understand one-sample hypothesis testing and p-values.**

One of the inferential statistical techniques that we can apply, thanks to the CLT, is one-sample hypothesis testing of the mean. Generally speaking, hypothesis testing consists of two hypotheses, the null hypothesis, called $H_0$, and the opposite to $H_0$—the alternative hypothesis, called $H_1$ or $H_a$. The null hypothesis for one-sample hypothesis testing of the mean tests whether the population mean is equal to, less than or equal to, or greater than or equal to a particular constant. An excellent analogy for hypothesis testing is our judicial system. The null hypothesis, $H_0$, is that you are innocent, and the alternative hypothesis, $H_1$, is that you are guilty.

Once the hypotheses are identified, the statistical test statistic is calculated, whether it is a Z or t value. (For simplicity's sake, in our discussion here we will say only the Z values, i.e, $Z_{calc}$ or $Z_{critical}$, when this does pertain to other tests—e.g., t, F, $\chi^2$.) This $Z_{calc}$ is compared to what we will call the critical Z, $Z_{critical}$. The $Z_{critical}$ value is based upon what is called a level of significance, called $\alpha$, which is usually equal to 0.10, 0.05, or 0.01. The level of significance can be viewed as the probability of making an error (or mistake), given that the $H_0$ is correct. Relating this to the judicial system, this is the probability of wrongly determining someone is guilty when in reality they are innocent, so we want to keep the level of significance rather small. Remember that statistics is not an exact science. We are dealing with estimates of the actual values. (The only way that we can be completely certain is if we use the entire population.) So, we want to keep the likelihood of making an error relatively small.

There are two possible statistical decisions and conclusions that are based on comparing the two Z values, $Z_{calc}$ and $Z_{critical}$. If $|Z_{calc}| > |Z_{critical}|$, we reject $H_0$. When we reject $H_0$, there is enough statistical evidence to support $H_1$. That is, in terms of the judicial system, there is overwhelming evidence to conclude that the individual is guilty. On the other hand, we do fail to reject $H_0$ when $|Z_{calc}| \leq |Z_{critical}|$, and we conclude that there is *not* enough statistical evidence to support $H_1$. That is, in terms of the judicial system, there is not enough evidence to conclude that the individual is guilty. The judicial system would then say that the person is innocent, but, in reality this is not necessarily true. We just did not have enough evidence to say that the person is guilty.

As we discussed under Fundamental Concept 3, understanding Z scores, the $|Z_{critical}|$ and also the $|Z_{calc}|$ are not simply numbers; they represent the number of standard deviations away from the mean that a value is. In this case, it is the number of standard deviations away from the hypothesized value used in $H_0$. So, we reject $H_0$ when we have a relatively large $|Z_{calc}|$, i.e., $|Z_{calc}| > |Z_{critical}|$. That is, we reject $H_0$ when the value is a relatively large number of standard deviations away from the hypothesized value. Conversely, when we have a relatively small $|Z_{calc}|$ (i.e., $|Z_{calc}| \leq |Z_{critical}|$ ), we fail to reject $H_0$. That is, the $|Z_{calc}|$ value is relatively near the hypothesized value and could be simply due to random variation.

Instead of comparing the two Z values, $Z_{calc}$ and $Z_{critical}$, another more generalizable approach that can also be used with other hypothesis tests (i.e., tests involving other distributions like t, F, $\chi^2$) is a concept known as p-values. The p-value is the probability of rejecting $H_0$. Thus, in terms of the one-sample hypothesis test using the Z, the p-value is the probability that is associated with $Z_{calc}$. So, as shown in Table 2.1, a relatively large $|Z_{calc}|$ results in rejecting $H_0$ and has a relatively small p-value. Alternatively, a relatively small $|Z_{calc}|$ results in not rejecting $H_0$ and has a relatively large p-value.

**Table 2.1 Decisions and Conclusions to Hypothesis Tests in Relationship to the p-value**

| Critical Value | p-value | Statistical Decision | Conclusion |
|---|---|---|---|
| $\left\| Z_{CALC} \right\| > \left\| Z_{Critical} \right\|$ | p-value $< \alpha$ | Reject $H_0$ | There is enough evidence to say that $H_1$ is true. |
| $\left\| Z_{CALC} \right\| \leq \left\| Z_{Critical} \right\|$ | p-value $\geq \alpha$ | Do Not Reject $H_0$ | There is not enough evidence to say that $H_1$ is true. |

General interpretation of a p-value is if it is:

- Less than 1%, there is overwhelming evidence that supports the alternative hypothesis.
- Between 1% and 5%, there is strong evidence that supports the alternative hypothesis.
- Between 5% and 10%, there is weak evidence that supports the alternative hypothesis.
- Greater than 10%, there is little to no evidence that supports the alternative hypothesis.

An excellent real-world example of p-values is the criterion that the Federal Food and Drug Administration (FDA) uses to approve new drugs. A new drug is said to be effective if it has a p-value less than 0.05 (and FDA does not change the threshold of 0.05). So, a new drug is approved only if there is strong evidence that it is effective.

# Many Approaches/Techniques Are Correct, and a Few Are Wrong

**Fundamental Concept 6: Many approaches/techniques are correct, and a few are wrong.**

In your first statistics course, many and perhaps an overwhelming number of approaches and techniques were presented. When do you use them? Do you remember why you use them? Some approaches/techniques should not even be considered with some data. Two major questions should be asked when considering applying a statistical approach/technique:

1.  Is it statistically appropriate?

2.  What will it possibly tell us?

An important factor to consider in deciding which technique to use is whether one or more of the variables is categorical or continuous. Categorical data can be nominal data such as gender, or it may be ordinal such as the Likert scale. Continuous data can have fractions (or no fractions, in which the data is an integer), and we can measure the distance between values. But with categorical data, we cannot measure distance. Simply in terms of graphing, we would use bar and pie charts for categorical data but not for continuous data. On the other hand, graphing a continuous variable requires a histogram. When summarizing data, descriptive statistics are insightful for continuous variables; a frequency distribution is much more useful for categorical data.

To illustrate, we will use the data in Table 2.2 and in the file Countif.xls and worksheet rawdata. The data consists of survey data from 20 students asking them how useful their statistics class was (column C), where 1 represents extremely not useful and 5 represents extremely useful, along with some individual descriptors of major (Business or Arts and Sciences (A&S)), Gender, current salary, GPA, and years since graduating. Major and Gender (and correspondingly Gender code) are examples of nominal data. The Likert scale of usefulness is an example of ordinal data. Salary, GPA, and years are examples of continuous data.

**Table 2.2  Data and Descriptive Statistics in Countif.xls file and Worksheet Statistics**

| | A | B | C | D | E | F | G |
|---|---|---|---|---|---|---|---|
| 1 | Major | Gender | Usefulness | Salary | Gender code | GPA | Years |
| 2 | Business | Male | 3 | 52125 | 0 | 3.53 | 4.40 |
| 3 | Business | Female | 1 | 52325 | 1 | 2.58 | 4.18 |
| 4 | Business | Male | 4 | 63042 | 0 | 3.52 | 5.30 |
| 5 | A&S | Male | 3 | 54928 | 0 | 3 | 4.49 |
| 6 | Business | Male | 4 | 50599 | 0 | 3.22 | 4.06 |
| 7 | A&S | Female | 2 | 42036 | 1 | 3.06 | 3.87 |
| 8 | A&S | Female | 3 | 46427 | 1 | 2.35 | 4.64 |
| 9 | A&S | Male | 3 | 51865 | 0 | 3.22 | 5.08 |
| 10 | A&S | Female | | 33263 | 1 | 2.86 | 2.03 |
| 11 | Business | Female | 5 | 58434 | 1 | 3.6 | 5.76 |
| 12 | Business | Male | 4 | 61551 | 0 | 3 | 5.38 |
| 13 | A&S | Male | 5 | 31235 | 0 | 3.11 | 1.38 |
| 14 | Business | Male | 2 | 58730 | 0 | 3.43 | 4.83 |
| 15 | A&S | Female | 4 | 35830 | 1 | 3.31 | 2.91 |
| 16 | Business | Male | | 53267 | 0 | 2.62 | 4.87 |
| 17 | Business | Male | 5 | 65437 | 0 | 3.28 | 5.54 |
| 18 | A&S | Female | 4 | 47591 | 1 | 2.68 | 3.76 |
| 19 | A&S | Female | 4 | 42659 | 1 | 3.16 | 3.27 |
| 20 | Business | Male | 3 | 50996 | 0 | 3.84 | 4.33 |
| 21 | A&S | Male | 2 | 40185 | 0 | 3.29 | 3.02 |
| 22 | A&S | Male | 5 | 33155 | 0 | 3.69 | 2.32 |
| 23 | Business | Female | 1 | 52695 | 1 | 2.54 | 3.38 |
| 24 | | | | | | | |
| 25 | min | | 1 | 31235 | 0 | 2.35 | 1.38 |
| 26 | max | | 5 | 65437 | 1 | 3.84 | 5.76 |
| 27 | average | | 3.35 | 49017.05 | 0.409090909 | 3.131364 | 4.036364 |
| 28 | median | | 3.5 | 51430.5 | 0 | 3.19 | 4.255 |
| 29 | stdev | | 1.2680279 | 9937.653 | 0.50323628 | 0.400705 | 1.18139 |

Using some Excel functions, we provided some descriptive statistics in rows 25 to 29 in the stats worksheet. These descriptive statistics are valuable in understanding the continuous data—e.g., the fact that the average is less than the median, and the salary data is slightly left-skewed with a minimum of $31,235, a maximum of $65,437 and a mean of $49,017. Descriptive statistics for the categorical data are not very helpful. E.g., for the usefulness variable, an average of 3.35 was calculated, slightly above the middle value of 3. A frequency distribution would give much more insight.

Use JMP and open Countif.xls file. There are two ways that you can open an Excel file. One way is similar to opening any file in JMP, and the other way is directly from inside Excel (when JMP has been added to Excel as an Add-in).

1. To open the file in JMP, first click to open JMP. From the top menu, click **File→Open**. Locate the Countif.xls Excel file on your computer and click on it in the selection window. You get the Open Data File dialog box. Click on option Best Guess under Should Row 1 be Labels? Check the box for Allow individual worksheet selection. Click **OPEN**. The Select the sheets to import dialog box with the file's worksheets listed will appear. Click rawdata and click **OK**. The data table should then appear.

2. If you want to open JMP from within Excel (and you can be in stats or rawdata), on the top Excel menu click **JMP**. (Note: The first time you use this approach, select **Preferences**. Check the box for Use the first row s as column names. Click **OK**. Subsequent use of this approach does not require you to click **Preferences**.) Highlight cells A1:G23. Click Data Table. JMP should open and the data table will appear.

### Figure 2.7 Modeling Types of Gender

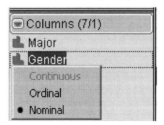

In JMP, as illustrated in Figure 2.7, move your cursor to the Columns panel and the red bar chart symbol, next to the variable Gender. When your pointer changes to a hand, then right-click. You will get three rows of options: continuous (which is grayed out), ordinal, and nominal. Next to nominal will be a dark colored dot, which indicates JMP's best guess of what type of data the column Gender is Nominal.

If you move your cursor over the blue triangle, [▲], beside Usefulness, you will see the black dot next to Continuous. But actually the data is ordinal. So click Ordinal. JMP now considers that column as ordinal (note that the blue triangle changed to green bars, [▲]).

Following the same process, change the column Gender code to nominal (the blue triangle now changes to red bars). The Table should look like Figure 2.8. To save the file as a JMP file, first, in the Table panel, right-click **Notes** and select **Delete**. At the top menu, click **File→Save As**, type as filename Countif, and click **OK**.

## Figure 2.8  Countif.jmp after Modeling Type Changes

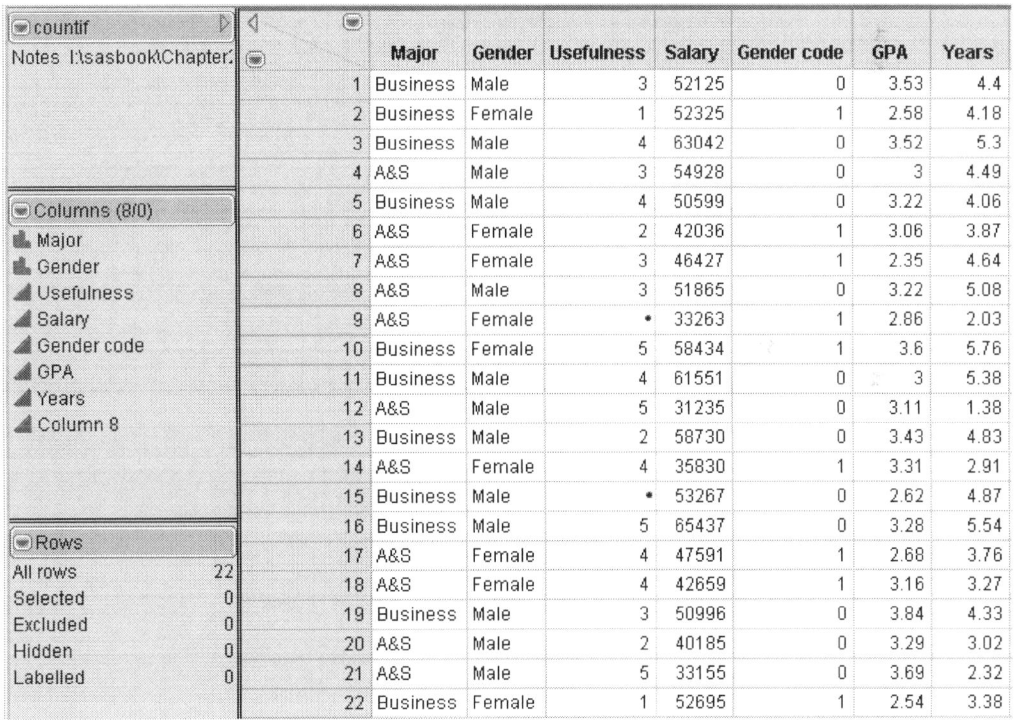

| | Major | Gender | Usefulness | Salary | Gender code | GPA | Years |
|---|---|---|---|---|---|---|---|
| 1 | Business | Male | 3 | 52125 | 0 | 3.53 | 4.4 |
| 2 | Business | Female | 1 | 52325 | 1 | 2.58 | 4.18 |
| 3 | Business | Male | 4 | 63042 | 0 | 3.52 | 5.3 |
| 4 | A&S | Male | 3 | 54928 | 0 | 3 | 4.49 |
| 5 | Business | Male | 4 | 50599 | 0 | 3.22 | 4.06 |
| 6 | A&S | Female | 2 | 42036 | 1 | 3.06 | 3.87 |
| 7 | A&S | Female | 3 | 46427 | 1 | 2.35 | 4.64 |
| 8 | A&S | Male | 3 | 51865 | 0 | 3.22 | 5.08 |
| 9 | A&S | Female | • | 33263 | 1 | 2.86 | 2.03 |
| 10 | Business | Female | 5 | 58434 | 1 | 3.6 | 5.76 |
| 11 | Business | Male | 4 | 61551 | 0 | 3 | 5.38 |
| 12 | A&S | Male | 5 | 31235 | 0 | 3.11 | 1.38 |
| 13 | Business | Male | 2 | 58730 | 0 | 3.43 | 4.83 |
| 14 | A&S | Female | 4 | 35830 | 1 | 3.31 | 2.91 |
| 15 | Business | Male | • | 53267 | 0 | 2.62 | 4.87 |
| 16 | Business | Male | 5 | 65437 | 0 | 3.28 | 5.54 |
| 17 | A&S | Female | 4 | 47591 | 1 | 2.68 | 3.76 |
| 18 | A&S | Female | 4 | 42659 | 1 | 3.16 | 3.27 |
| 19 | Business | Male | 3 | 50996 | 0 | 3.84 | 4.33 |
| 20 | A&S | Male | 2 | 40185 | 0 | 3.29 | 3.02 |
| 21 | A&S | Male | 5 | 33155 | 0 | 3.69 | 2.32 |
| 22 | Business | Female | 1 | 52695 | 1 | 2.54 | 3.38 |

Table panel (left side):
- countif
- Notes I:\sasbook\Chapter:
- Columns (8/0): Major, Gender, Usefulness, Salary, Gender code, GPA, Years, Column 8
- Rows: All rows 22, Selected 0, Excluded 0, Hidden 0, Labelled 0

At the top menu in JMP, select **Analyze→Distribution**. The Distribution dialog box will appear. In this new dialog box, click Major, hold down the shift key, click Years, and release. All the variables should be highlighted, as in Figure 2.9.

**Figure 2.9  The JMP Distribution Dialog Box**

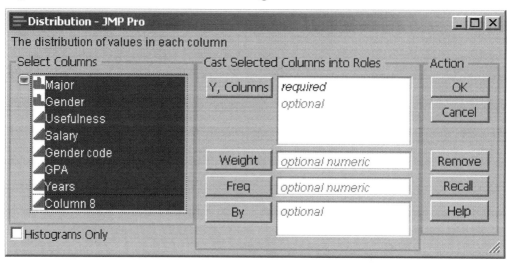

Click **Y, Columns**, and all the variables should be transferred over to the box to the right. Click **OK** and a new window will appear. Examine Figure 2.10 and your Distribution window in JMP. All the categorical variables (**Major**, **Gender**, **Usefulness**, and **Gender code**), whether they are nominal or ordinal, have frequency numbers and a histogram, not descriptive statistics. But the continuous variables have descriptive statistics and a histogram.

As shown in Figure 2.10, click the area/bar of the Major histogram for Business. You can immediately see the distribution of Business students; they are highlighted in each of the histograms.

## Figure 2.10  Distribution Output for Countif.xls Data

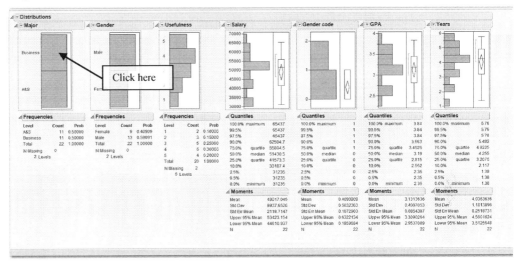

What if you want to further examine the relationship between Business and these other variables or the relationship between any two of these variables (i.e., perform some bivariate analysis). You can click any of the bars in the histograms to see the corresponding data in the other histograms. We could possibly look at every combination, but what is the right approach?

JMP provides excellent direction. Select **Analyze→Fit Y by X**. The bivariate diagram in the lower left of the new window, as in Figure 2.11, provides guidance on which technique is appropriate. For example, select **Analyze→Fit Y by X**. Drag Salary to the white box to the right of **Y, Response** (or click Salary and then click **Y, Response**). Similarly, click **Years**, hold down the left mouse button, and drag it to the white box to the right of **X, Factor**. The Fit Y by X dialog box should look like Figure 2.11. According to the lower left diagram in Figure 2.11, bivariate analysis will be performed. Click **OK.**

**Figure 2.11  Fit Y by X Dialog Box**

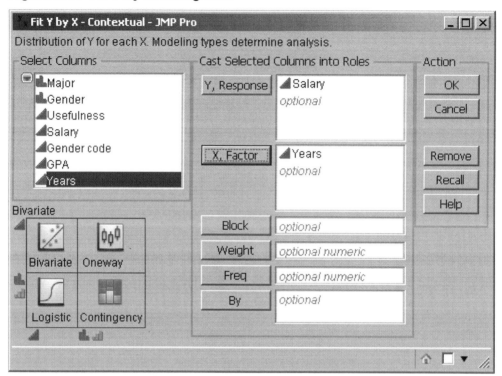

Click the red triangle in the **Bivariate Fit of Salary by Years** window, and click **Fit Line**. The output will look like Figure 2.12. The large positive coefficient of 7743.7163 demonstrates a strong positive relationship. (Positive implies that as **Years** increases **Salary** also increases; or the slope is positive. In contrast, a negative relationship has a negative slope. So, as the X variable increases, the Y variable decreases.) The RSquare value or the coefficient of determination is 0.847457, which also shows a strong relationship.

**Figure 2.12  Bivariate Analysis of Salary by Years**

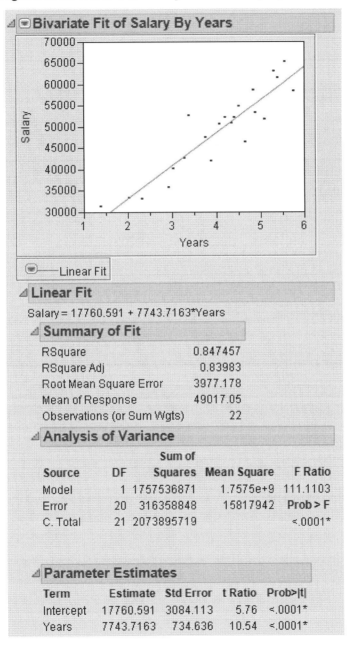

⊿ ⊡ Bivariate Fit of Salary By Years

⊡——Linear Fit

⊿ **Linear Fit**

Salary = 17760.591 + 7743.7163*Years

⊿ **Summary of Fit**

| | |
|---|---|
| RSquare | 0.847457 |
| RSquare Adj | 0.83983 |
| Root Mean Square Error | 3977.178 |
| Mean of Response | 49017.05 |
| Observations (or Sum Wgts) | 22 |

⊿ **Analysis of Variance**

| Source | DF | Sum of Squares | Mean Square | F Ratio |
|---|---|---|---|---|
| Model | 1 | 1757536871 | 1.7575e+9 | 111.1103 |
| Error | 20 | 316358848 | 15817942 | Prob > F |
| C. Total | 21 | 2073895719 | | <.0001* |

⊿ **Parameter Estimates**

| Term | Estimate | Std Error | t Ratio | Prob>|t| |
|---|---|---|---|---|
| Intercept | 17760.591 | 3084.113 | 5.76 | <.0001* |
| Years | 7743.7163 | 734.636 | 10.54 | <.0001* |

RSquare values can range from 0 (no relationship) to 1 (exact/perfect relationship). The square root of the RSquare multiplied by 1 if it has a positive slope (that is, for this example the coefficient for years is positive) or multiplied by -1 if it has a negative slope is equal to the correlation. Correlation values near -1 or 1 show strong relationships. (A negative correlation implies a negative relationship, while a positive correlation implies a positive relationship.) Correlation values near 0 imply no linear relationship.

On the other hand, what if we drag **Major** and **Gender** to the **Y, Response** and **X, Factor**, respectively, and click **OK**. The bivariate analysis diagram on the lower left in Figure 2.11 would suggest a Contingency analysis. The contingency analysis is shown in Figure 2.13.

The Mosaic Plot visually graphs the percentages from the contingency table. From the Mosaic plot, visually there appears to be a significant difference in **Gender** by **Major.** However, looking at the $\chi^2$ test of independence results, the p-value or Prob>ChiSq is 0.1933. The $\chi^2$ test assesses whether the row variable is significantly related to the column variable. That is, in this case, is **Gender** related to **Major** and vice versa? With a p-value of 0.1993, we would fail to reject $H_0$ and conclude that there is not a significant relationship between **Major** and **Gender**. (It should be noted that performing the $\chi^2$ test with this data is not advised because some of the expected values are less than 5.)

**Figure 2.13  Contingency Analysis of Major by Gender**

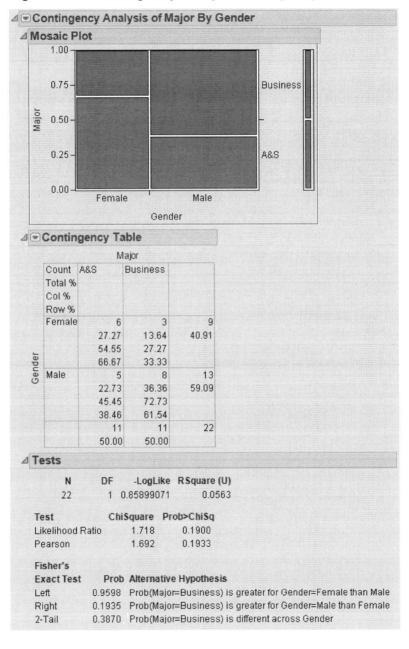

Contingency Analysis of Major By Gender

**Mosaic Plot**

**Contingency Table**

Major

| Count<br>Total %<br>Col %<br>Row % | A&S | Business | |
|---|---|---|---|
| Female | 6 | 3 | 9 |
| | 27.27 | 13.64 | 40.91 |
| | 54.55 | 27.27 | |
| | 66.67 | 33.33 | |
| Male | 5 | 8 | 13 |
| | 22.73 | 36.36 | 59.09 |
| | 45.45 | 72.73 | |
| | 38.46 | 61.54 | |
| | 11 | 11 | 22 |
| | 50.00 | 50.00 | |

**Tests**

| N | DF | -LogLike | RSquare (U) |
|---|---|---|---|
| 22 | 1 | 0.85899071 | 0.0563 |

| Test | ChiSquare | Prob>ChiSq |
|---|---|---|
| Likelihood Ratio | 1.718 | 0.1900 |
| Pearson | 1.692 | 0.1933 |

**Fisher's Exact Test**

| | Prob | Alternative Hypothesis |
|---|---|---|
| Left | 0.9598 | Prob(Major=Business) is greater for Gender=Female than Male |
| Right | 0.1935 | Prob(Major=Business) is greater for Gender=Male than Female |
| 2-Tail | 0.3870 | Prob(Major=Business) is different across Gender |

As we illustrated, JMP, in the bivariate analysis diagram of the Fit Y by X dialog box, helps the analyst select the proper statistical method to use. The Y variable is usually considered to be a dependent variable. For example, if the X variable is continuous and the Y is categorical (nominal or ordinal), then in the lower left of the diagram in Figure 2.11 logistic regression will be used. This will be discussed in Chapter 5. In another scenario, with the X variable as categorical and the Y variable as continuous, JMP will suggest One-way ANOVA, which will also be discussed in Chapter 4. If there is no dependence in the variables of interest, as you will learn in this book, there are other techniques to use.

Additionally, depending on the type of data, some techniques are appropriate and some are not. As we can see, one of the major factors is the type of data being considered—i.e., continuous or categorical. While JMP is a great help, just because an approach/technique appears appropriate, before running it, you need to step back and ask yourself what the results could provide. Part of that answer requires understanding and having knowledge of the problem situation. For example, we could be considering the bivariate analysis of GPA and Years. But, logically they are not related, and if a relationship is demonstrated it would most likely be a spurious one. What would it mean?

So you may decide you have an appropriate approach/technique, and it could provide some meaningful insight. However, we cannot guarantee that you will get the results you expect or anticipate. We are not sure how it will work out. Yes, the approach/technique is appropriate. But depending on the theoretical and actual relationship that underlies the data, it may or may not be helpful. This process of exploration is all part of developing and telling the statistical story behind the data.

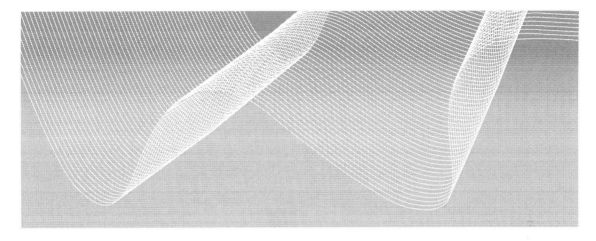

# Chapter 3

## Introduction to Multivariate Data

## Multivariate Data and Multivariate Data Analysis

Most data sets in the real-world are multivariate; that is, they contain more than two variables. Generally speaking, a data set can be viewed conceptually as shown in Figure 3.1, where the columns represent variables and the rows are objects. The rows or objects are the entities by which the observations/measurements were taken (for example, by event, case, transaction, person, companies, etc.). The columns or variables are the characteristics by which the objects were measured. Multivariate data analysis is when more than two variables are analyzed simultaneously.

## Figure 3.1  A Conceptual View of a Data Set

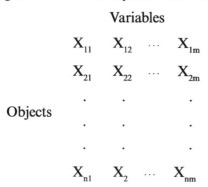

Figure 3.2 provides a framework of statistical and visual methods for analyzing multivariate data. The initial multivariate analytical steps should be the data discovery of relationships through tables and graphs. Some of this data discovery will include univariate descriptive statistics or distribution analysis and perhaps some bivariate analyses (e.g., scatterplots and contingency tables, as discussed in Chapter 2). In this chapter, we will explore some of the multivariate tables and graphs to consider.

**Figure 3.2 A Framework to Multivariate Analysis**

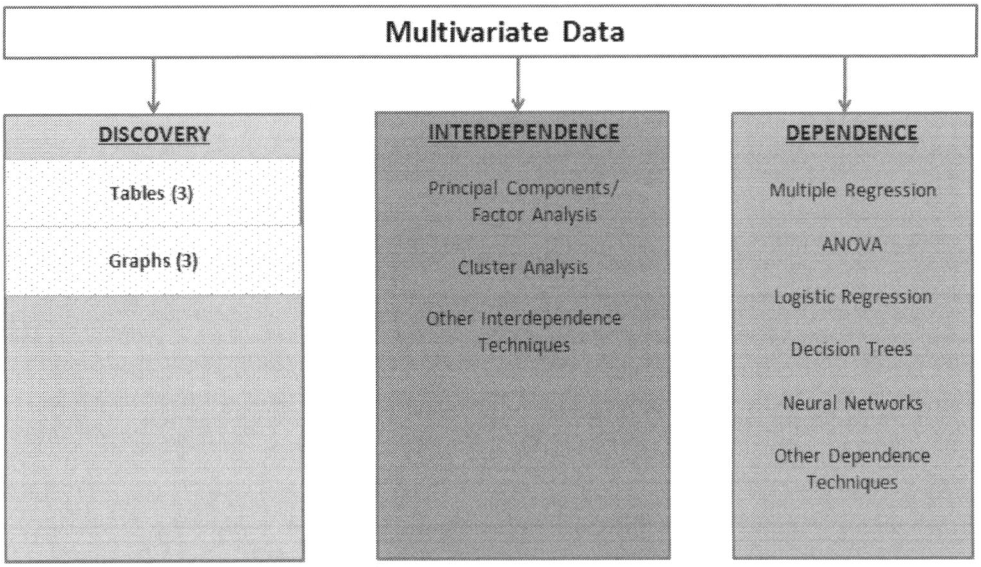

As shown in Figure 3.2, there are numerous multivariate statistical and data mining techniques available to analyze multivariate data. Several of the more popular multivariate, and a few of the data mining, techniques will be discussed in this text (factor analysis/principal components, cluster analysis, multiple regression, ANOVA, logistic regression, and decision trees). Generally speaking, we categorize these techniques into interdependence and dependence techniques, as shown in Figure 3.2. The interdependence techniques examine relationships either between the variables (columns) or the observations (rows) without considering causality. With dependence techniques, one or more variables are identified as the dependent variable(s), and one assumes X variables cause Y(s). The objective of these dependence techniques is to examine and measure the relationship between other (independent) variables and the dependent variable(s). Two dependence techniques that may have been covered in an introductory statistics course, multiple regression and ANOVA, will be reviewed in Chapter 4.

One last word of caution/advice before we start our multivariate journey (this is just a reiteration of the sixth fundamental concept discussed in Chapter 2). With all the many discovery tools and the numerous statistical techniques, we cannot guarantee that they will produce any useful insights until you try them. Successful results really depend on the data set. But, on the other hand, there are times when it is inappropriate to use a tool or technique. Throughout the following chapters, we will provide some guidance of when or when not to use a tool/technique.

## Using Tables to Explore Multivariate Data

Row and Column tables, contingency tables, crosstabs, Excel PivotTables, and JMP Tabulate are basic OLAP tools that are used to produce tables to examine/summarize relationships between two or more variables. To illustrate, we will use the survey data from 20 students in the file Countif.xls and in the worksheet rawdata from Chapter 2, as shown in Table 2.2.

To generate a PivotTable in Excel:

1. Highlight cells A1:G23. Click **Insert** on the top menu. A new set of icons appears. All the way to the left, click PivotTable. When the dialog box appears, click **OK**.

2. A new worksheet will appear similar to Figure 3.3. In the upper part of the PivotTable Field List box (on the right side of the worksheet), drag Major all the way to the left, in column A where it says Drop Row Fields Here and release the mouse button. Similarly, drag Gender to where it says Drop Column Fields Here.

   Notice back in the upper part of the PivotTable Field List box, under Choose fields to add to report, that Major and Gender have check marks next to them. In the lower part of the box, under Column labels, see Gender; under Row labels see Major.

3. Drag Salary to the left where it says Drop Value Fields Here. Salary in the upper part of the PivotTable Field List box is now checked. Below under ∑ Values is the Sum of Salary. Click the drop arrow for the Summary of Salary to open a list of options. Click the Value Field Setting, and a new dialog box similar to Figure 3.4 will appear. In the dialog box, you can change the displayed results to various summary statistics. Click Average (notice that the Custom Name changes to Average of Salary). Click **OK.**

The resulting PivotTable should look similar to Figure 3.5, which shows the average salary by Gender and Major.

## Figure 3.3  PivotTable Worksheet

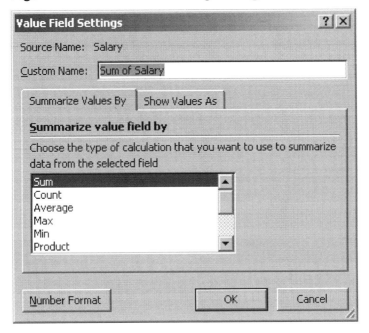

## Figure 3.4  Value Field Setting Dialog Box

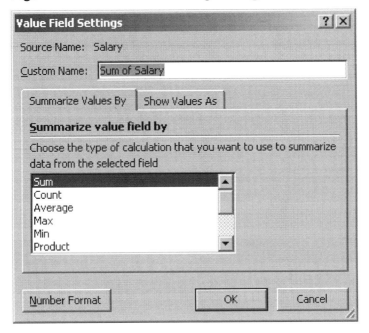

As the JMP diagram in the Fit Y by X dialog box directs us (as shown in Figure 2.10), the rows and columns are categorical data in a contingency table. This is, in general, the same for these OLAP tools. The data in the PivotTable can be further sliced by dragging a categorical variable into the **Drop Report Filter Fields Here** area.

## Figure 3.5  Resulting Excel PivotTable

| | A | B | C | D |
|---|---|---|---|---|
| 1 | | | | |
| 2 | | | | |
| 3 | Average of Salary | Gender ▾ | | |
| 4 | Major ▾ | Female | Male | Grand Total |
| 5 | A&S | 41301 | 42273.6 | 41743.09091 |
| 6 | Business | 54484.66667 | 56968.375 | 56291 |
| 7 | Grand Total | 45695.55556 | 51316.53846 | 49017.04545 |

To generate a similar table in JMP:

1. Open **Countif.jmp**. Click **Tables→Tabulate**.  The Tabulate dialog box appears (Figure 3.6).

**Figure 3.6 The Tabulate Dialog Box**

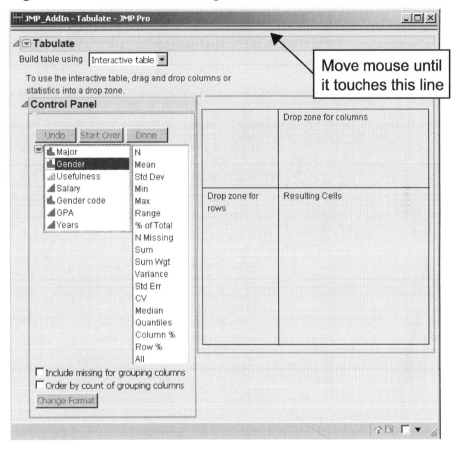

2. Drag **Salary** to the **Resulting Cells** area. The table now shows the sum of all the salaries (1078375). In the Control Panel, drag **Mean** to Sum (1078375). The average salary is 49017.0454. Drag **Gender** to **Salary**, where it now shows Salary (Female Mean=45695.555556 and Male Mean=51316.538462). Lastly, drag **Major** to the white square to the left of the Mean for Females (45695.55). Click the red triangle and click **Show Chart**. The table and chart should now look like Figure 3.7.

3. If you want to copy and paste the JMP table and chart into Microsoft Word or Microsoft PowerPoint, as shown in Figure 3.6, move your mouse to the top horizontal lines near the top of the Tabulate dialog box. The result is the activation of the Menu and Toolbars, as seen at the top of Figure 3.7. A row of icons will appear as shown in Figure 3.7. Click the Selection icon, as shown

in Figure 3.7. Click in the table, hold down the shift key, and click in the chart. Right-click the mouse and select **Copy**. Go to a Word or PowerPoint file and paste them in. The table and chart will appear like Figure 3.8. Note: To copy and paste JMP results from other JMP tools/techniques, you follow this same process.

**Figure 3.7  Selection Button to Copy the Output**

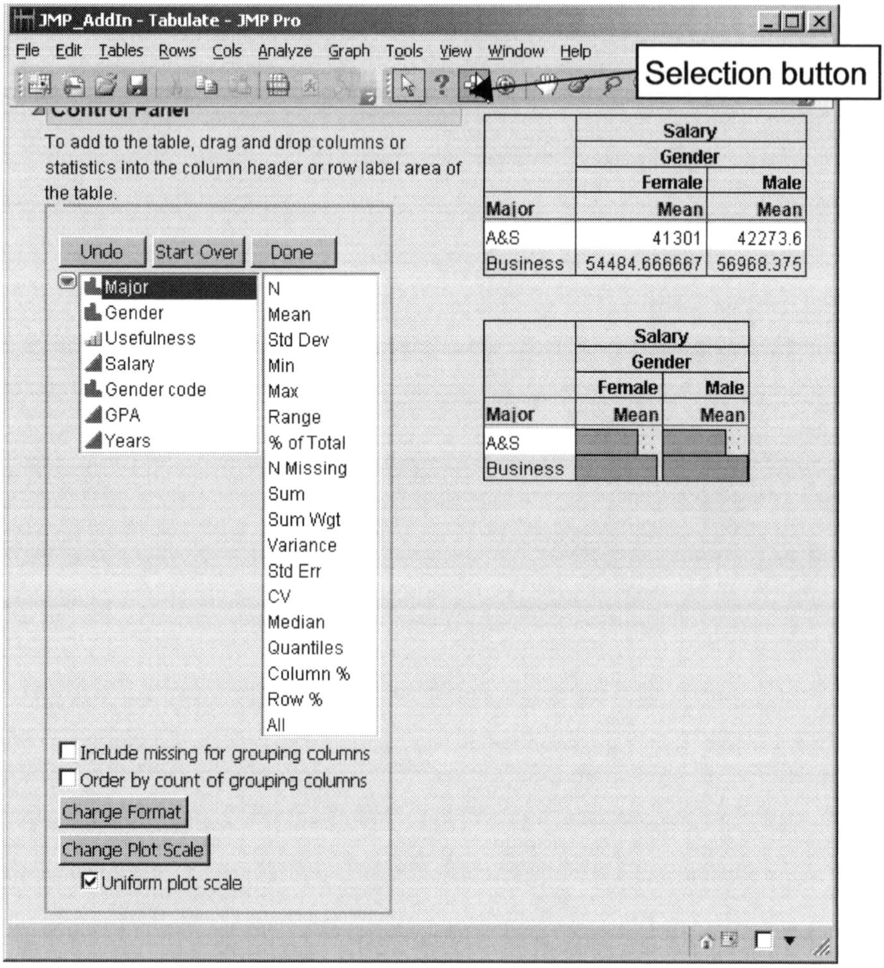

**Figure 3.8  Resulting Copy of the Tabulate Table and Chart (into a Microsoft Word Document)**

|  | Salary | |
|---|---|---|
|  | Gender | |
|  | Female | Male |
| Major | Mean | Mean |
| A&S | 41301 | 42273.6 |
| Business | 54484.666667 | 56968.375 |

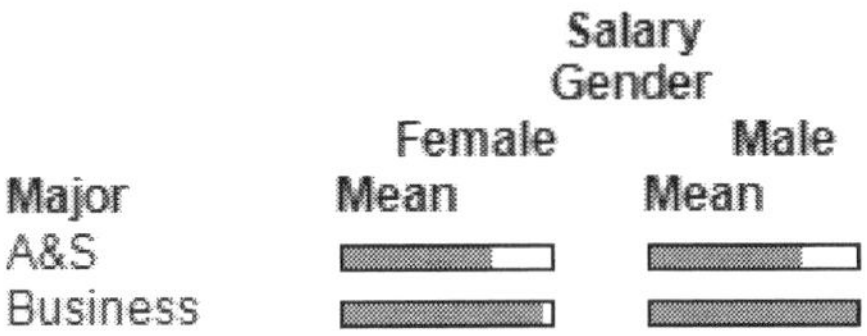

|  | Salary | |
|---|---|---|
|  | Gender | |
|  | Female | Male |
| Major | Mean | Mean |
| A&S |  |  |
| Business |  |  |

# Using Graphs to Explore Multivariate Data

Graphical presentation tools provide powerful insight into multivariate data. JMP provides a large selection of visualization tools beyond the simple XY scatterplots, bar charts, and pie charts. One such tool that is extremely useful in exploring multivariate data is the Graph Builder.

Let's initially look at and use the Graph Builder with the Countif.jmp data set.

Select **Graph→Graph Builder**. The Graph Builder dialog box will appear, as in Figure 3.9. On the left of the dialog box, there is a window that has a list of our variables, and to the right is a graphical sandbox in which we can design our graph. As noted in Figure 3.9, there are several areas called drop zones where we can drop the variables. We can bring over to these drop zones either continuous or categorical (ordinal and nominal) data.

## Figure 3.9 Graph Builder Dialog Box

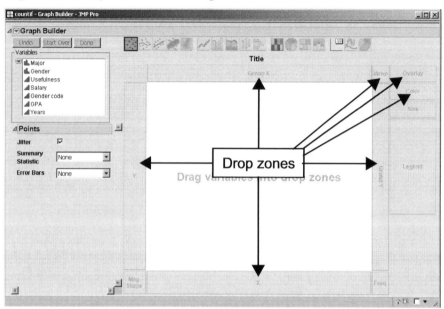

Click **Salary** and drag it around to the different drop zones. The graph immediately reacts as you move over the drop zones. Release **Salary** in the Y drop zone area. Now, click **Years** and again drag it around to the different drop zones and observe how the graph reacts, as shown in Figure 3.10. Notice that when you move **Years** to:

> **Group X** (you get several vertical graphs and further notice that years are grouped—or put into intervals).

> **Group Y** (you will get several horizontal graphs).

> **Wrap** and **Overlay** (**Years** is grouped into Year categories).

**Figure 3.10 Graph of Salary and Year when Year Is in Different Drop Zones**

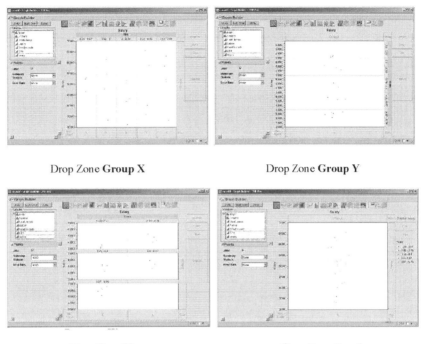

Drop Zone **Group X**     Drop Zone **Group Y**

Drop Zone **Wrap**     Drop Zone **Overlap**

Release Years in the X drop zone. Click Major and drag it around to the drop zones. Also, move Major over the Y and X drop zones, and observe how you can add another variable to those drop zones. Finally, release Major in the Group X drop zone. Lastly, click on Gender and drag it to the Group Y drop zone. There should now be four graphs as shown in Figure 3.11. We can quickly see that the Business students have higher salaries and are older (more years since they graduated).

### Figure 3.11 Graph of Salary versus Year by Major and Gender

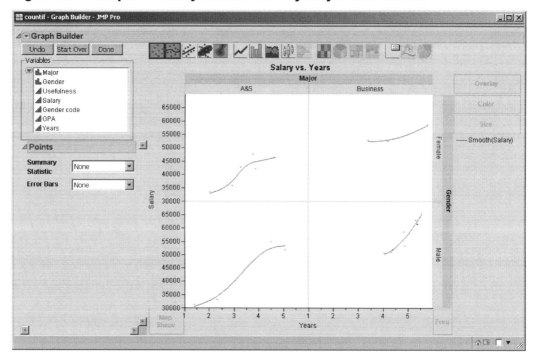

Another useful visualization tool for providing insights to multivariate data is the scatterplot matrix. Data that consists of k variables would require k(k-1) two-dimensional XY scatterplots; thus every combination of two variables is in an XY scatterplot. The upper triangle of graphs is the same as the lower triangle of graphs, except that the coordinates are reversed. Thus usually only one side of the triangle is shown. An advantage of the scatterplot matrix is that changes in one dimension can be observed by scanning across all the graphs in a particular row or column.

For example, let's produce a scatterplot matrix of the Countif.jmp data.

1.  Select **Graph→Scatterplot Matrix**, and the scatterplot matrix dialog box will appear. Hold down the **CTRL** key and click every variable one by one except for Gender Code. All the variables are now highlighted except for Gender Code. Click **Y, Columns** and all the variables are listed. Notice toward the lower left of the dialog box, there is Matrix Format; there is a drop-down arrow with Lower Triangular selected. If you click the drop-down arrow, you can see the different options available. Leave it as Lower Triangular. Click **OK**.

**Figure 3.12 Scatterplot Matrix of Countif.jmp**

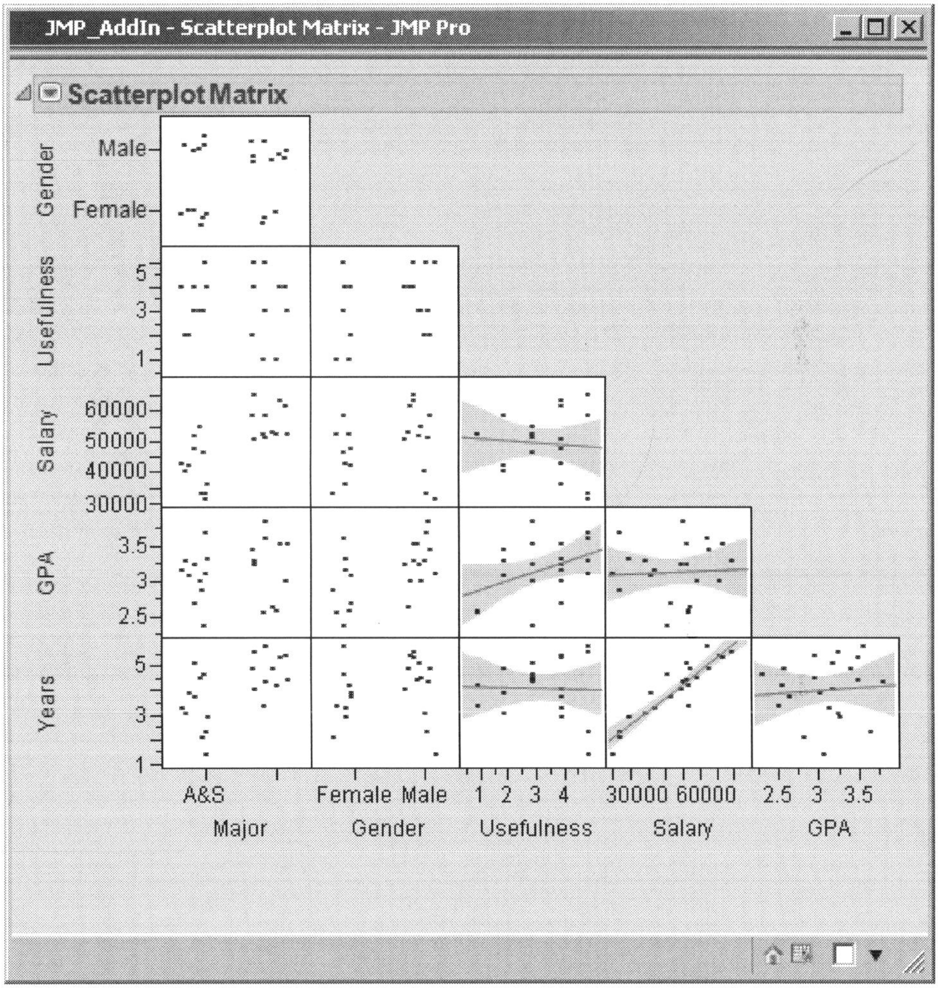

2.  A scatterplot matrix similar to Figure 3.12 will appear. For every scatterplot in the left-most column of the scatterplot matrix (which is labeled Major), the X axis is Major. Move the mouse on top of any point and left-click the mouse. That observation is now highlighted in that scatterplot but also in the other scatterplots. So you can observe how that observation performs in the other dimensions. Additionally, the row number appears. Similarly, you can hold down the left mouse button and highlight an area of points. The corresponding selected points in the other scatterplots are also highlighted as well as those observations in the data table.

3. Click the red triangle and Fit Line. In the scatterplots where both variables are continuous (in this case, Salary vs Years, Salary vs GPA, and GPA vs Years), a line is fitted with a confidence interval as shown in Figure 3.12. The smaller the width of the interval, the stronger the relationship. That is, Salary and Years have a strong relationship.

Generally speaking, the scatterplots in a scatterplot matrix are more informative with continuous variables as opposed to categorical variables. Nevertheless, insight into the effect of a categorical variable on the other variables can be visualized with the scatterplot matrix by using the Group option. For example:

1. Select **Graph→Scatterplot Matrix**. Click **Recall**, and all our variables are now listed under **Y, Columns**. Select Major→Group. Click **OK**.

2. The scatterplot matrix appears. Click the red triangle and **Shaded Ellipses**.

As shown in Figure 3.13, when we examine the ellipses, it appears that Arts and Sciences (A&S) students have lower salaries than Business students, and in the data set there were more A&S students that have recently graduated than Business students.

**Figure 3.13  Scatterplot with Shaded Ellipses**

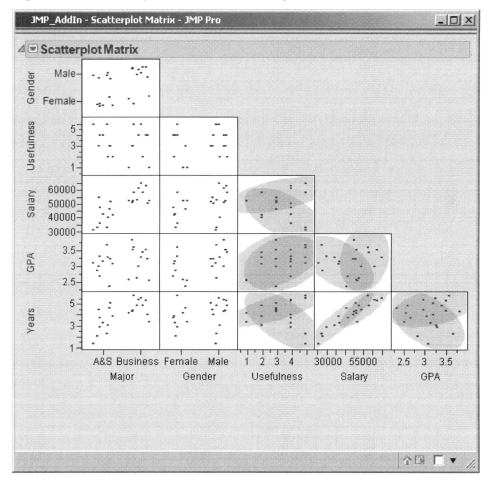

Let's explore another data set that contains time as a variable, the HomePriceIndexCPI1.jmp[1]. This file includes the quarterly home price index and consumer price index for each of the states in the United States and for the District of Columbia, from May 1975 until September 2009. This table has been sorted by State and Date. Initially, you could examine all the variables in a scatterplot matrix. If you do, the scatterplots are not very helpful—too many dates and states. Let's use the Graph Builder:

> In JMP, open HomePriceIndexCPI1.jmp. Select **Graph**→**Graph Builder**. Click Date and drag it to the X drop zone. Click **Home Price Index**, and drag it to the Y drop zone. The graph is not too informative yet. Now, click **State** and drag it to the Wrap drop zone.

We now have 51 small charts, by State, of Home Price Index vs Date, as shown in Figure 3.14. This set of charts is called a trellis chart: for each level of a categorical variable, a chart is created (similarly, the charts in Figure 3.11 are a trellis chart). Trellis charts can be extremely effective in discovering relationships with multivariate data. With this data, we can see by state the trend of their home price index. Some states increase significantly, such as CA, HI, and DC (highlighted in red), and some very slowly, such as GA, MO, MS, and WV (highlighted in green).

**Figure 3.14  Trellis Chart of the Home Price Indices by State**

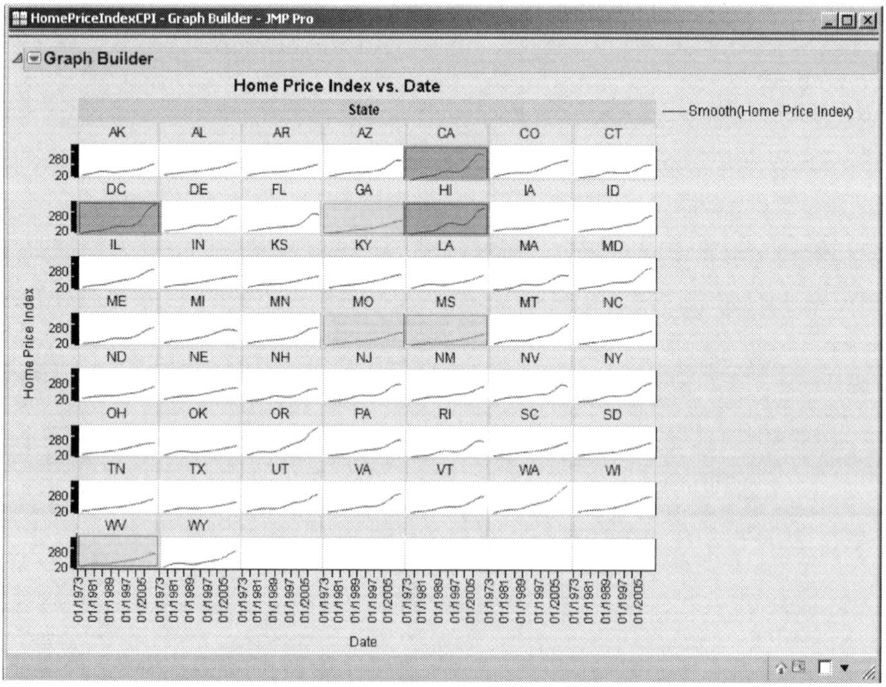

We can add CPI to the charts by clicking CPI and dragging it to the Y drop zone (just to the right of the Y axis label Home Price Index), as shown in Figure 3.15.

**Figure 3.15  Trellis Chart of Home Price Index and CPI by State**

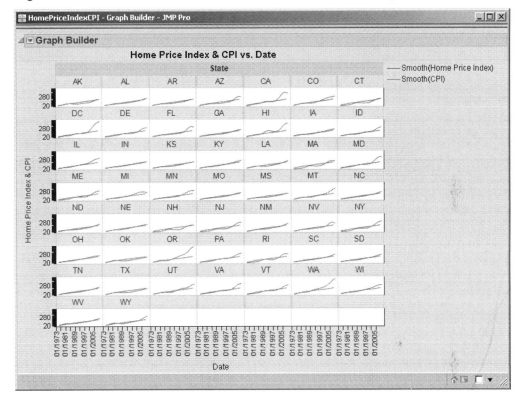

Another informative visualization chart is the Bubble plot, which can be very effective with data over time. To generate a Bubble plot with the Home Price Index data:

Select **Graph**→**Bubble Plot**. Drag Home Price Index to Y, drag State to X, drag Date to Time, and drag Home Price Index to Coloring, as shown in Figure 3.16. Click **OK**.

Click the go icon ![go icon], and you can observe how the home price indexes for the states increase over time until 2008, and then they decrease. Figure 3.17 captures the bubble plot at four dates. You can increase the speed by moving the slider to the right. Further, you can save this as a Flash file, so that you can send it in an e-mail or include it as part of a presentation, by clicking the red triangle and selecting **Save for Adobe Flash platform (.SWF)**.

**Figure 3.16  Bubble Plot Dialog Box**

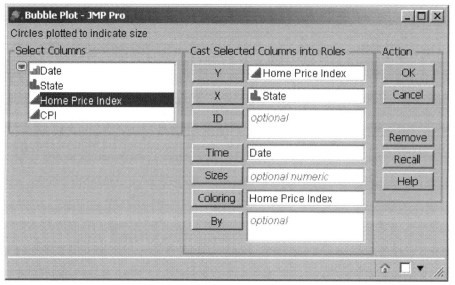

**Figure 3.17  Bubble Plot of Home Price Index by State for a Few Dates**

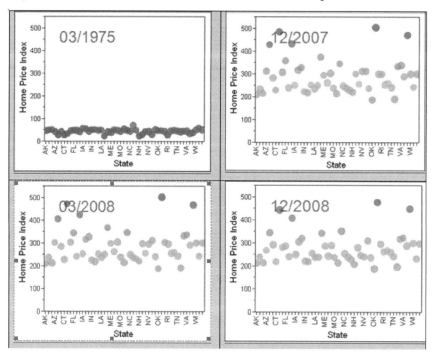

Let's examine another data set, profit by product.jmp[2]. This file contains sales data, including Revenue, Cost of Sales, and two calculated columns—Gross Profit and GP% (percentage of gross profit). (Notice that the ✛ to the right of the Gross Profit and GP% variables, respectively. Double-click the plus sign for either variable, and the Formula dialog box will appear with the corresponding formulas.)

The data is organized by time (quarter), distribution channel, product line, and customer ID. If we were examining this data set "in the real-world," the first step would be to generate and examine the results of some univariate and bivariate analyses with descriptive statistics, scatterplots, and tables. Next, since the data set is multivariate, graphics are likely to provide significant insights. So, let's use the JMP Graph Builder:

1.  Select **Graph**→**Graph Builder**. Drag Quarter to the X drop zone. Drag Revenue to the Y drop zone. Instead of viewing the boxplot, let's look at a line of the average revenue by quarter by right-clicking any open space in the graph and then selecting **Box Plot**→**Change to**→**Contour**. The density Revenue values are displayed by Quarter, as in Figure 3.18. Now, right-click inside any of the shaded revenue density areas. Then select **Contour**→**Change to**→**Line**, as shown in Figure 3.18.

**Figure 3.18  Revenue by Quarter Using Contours and Line Graphs**

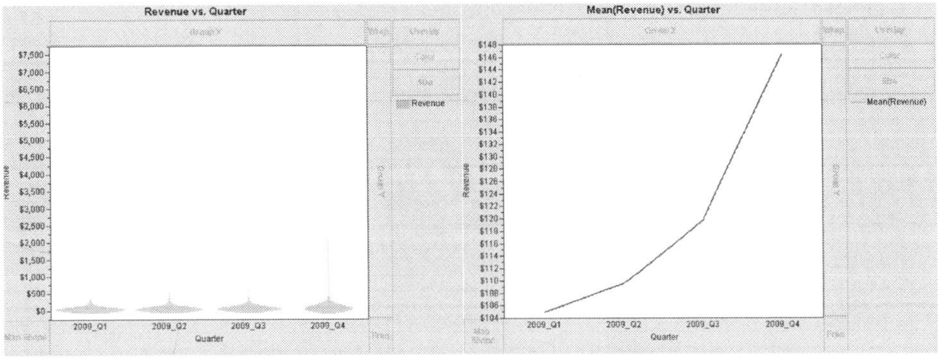

2.  Add Cost of Sales and Gross Profit to the Y axis, by dragging Cost of Sales to the Y drop zone. (Be careful to not remove Revenue—release the mouse button just to the right of the Y axis.) Similarly, drag Gross Profit just to the right of Y. Drag Product Line to the Wrap drop zone. Click Done. These steps should product a trellis chart similar to Figure 3.19.

**Figure 3.19  Trellis Chart of the Average Revenue, Cost of Sales, and Gross Product over Time (Quarter) by Product Line**

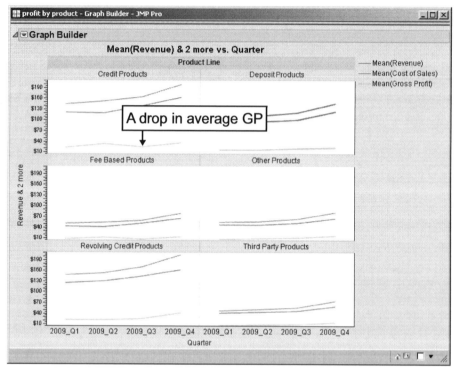

Examining Figure 3.19, we can see that, in terms of average Gross Profit, the Credit Products product line did not do too well in Quarter 3. We can focus or drill down on the Revenue and Cost of Sales of the Credit Products by using the Data Filter feature of JMP. The Data Filter when used in concert with the Graph Builder enhances our exploration capability.

3.  On the main menu in the data sheet window, select **Rows→Data Filter**. Click Product Line. Check the Include option. The graph now includes only Credit Products and looks like Figure 3.20. In the rows panel, note that 5,144 rows are Selected and 19,402 are Excluded.

**Figure 3.20  Graph of the Average Revenue, Cost of Sales and Gross Product by Quarter for the Credit Products Product Line**

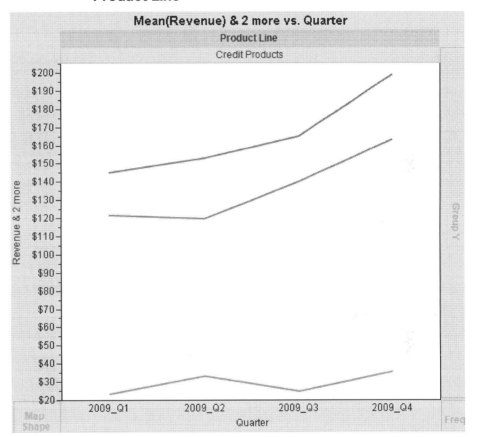

4.  Because we are looking only at Credit Products, drag Product Line from the top graph back over and release it on top of the other variables (i.e., at the left side in the Select Columns box). We can now focus on all our Credit Products customers by dragging Customer ID to the Wrap drop zone. The graph should look like Figure 3.21.

**Figure 3.21  Graph of the Average Revenue, Cost of Sales, and Gross Product by Quarter for the Credit Products Product Line by Customer ID**

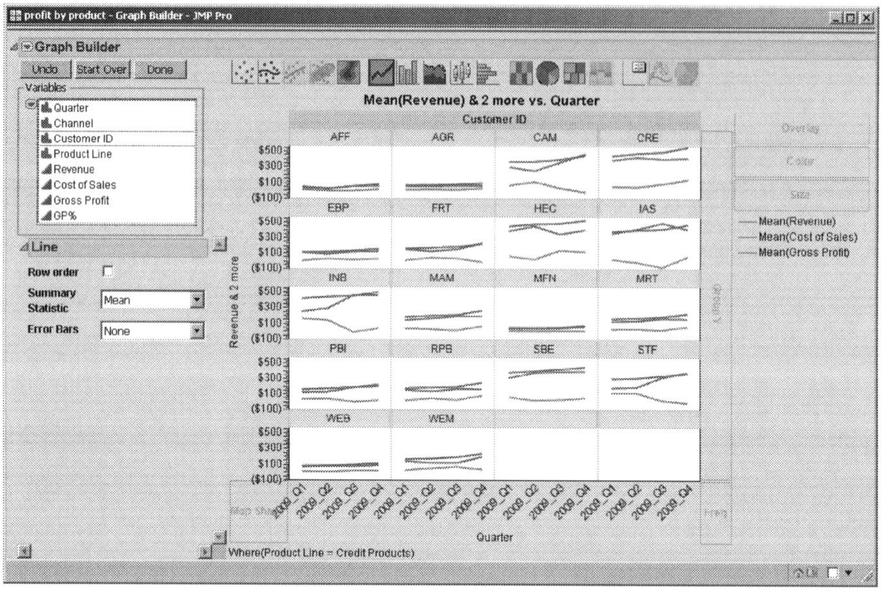

5.  However, if we have too many customers, this trellis chart may be way too busy. So, with the filter in play, let's build another graph. Click **Start Over** in the Graph Builder dialog box. Drag **Customer ID** to the Y drop zone; drag **Revenue** to the X drop zone. Right-click on any point and then select **Box Plot→Change to→Bar**. Drag **GP%** to the X drop zone. The graph should look similar to Figure 3.22

    We can quickly see for our Credit Products product line that the average sales (i.e., revenue) of customers IAS and CAM are relatively high, but their average percentage gross profit is rather low. We may want to further examine these customers separately. An effective feature of JMP is called dynamic linking: as you interact with the graph, you also interact with the data table. So, we can easily create a subset of the data with just these two customers.

6.  Click on one of the bars for **IAS**. Then hold down the shift key and click one of the bars for **CAM**. The bars for these two companies should be highlighted. In the Graph Builder window, click the View Associated Data icon [icon] in the extreme lower right.  The data table will be displayed with 64 rows selected. Select **Table→Subset**. In the Subset window, change the Output table name from its default to **CAM IAS profit by product**. Click **OK**. A new data table is created and displayed with only 64 rows for customers IAS and CAM only. Select

**File**→**Save**→**Save**. We can further analyze these two customers at a later time; let's return to Graph Builder.

**Figure 3.22  Bar Chart of the Average Revenue and %GP of Credit Products Customers**

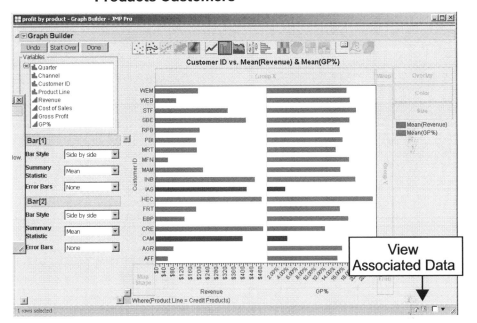

7.  If you have difficulty getting the Graph Builder dialog box back (Figure 3.22), navigate to the JMP Home Window and click profit by product—Graph Builder. Those two customers were rather easy to identify, but what if we had a large number of customers? We can sort the customers by average percent gross profit by redragging GP% to the Y drop zone. With the cursor somewhere on the Y axis, right-click and then click Ascending. Now, we can easily see the customers who are not doing well in terms of average percent gross profit as shown in Figure 3.23.

    Note, as annotated in Figure 3.22, if you click the small data table icon ▦ in the lower-right corner, a table will list the associated data with the graph.

**Figure 3.23  Bar Chart of the Average Revenue and %GP of Credit Products Customers in Ascending Order**

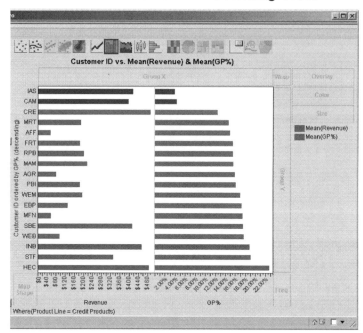

We can further explore the Credit Products product line and develop a better understanding of the data by adding another element to our filter. For example, we may postulate that there may be a correlation between average size of sales and gross profit. We can visualize this relationship by following this next step.

8.  In the Data Filter dialog box, click the **AND** icon. In the list of Add Filter Columns, click **Revenue**. Click **Add**. A scroll bar will appear.

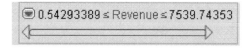

Double-click $0.54293389, change the value to 1350, and press **Enter**. As a result, 169 rows were matched. The graph should look similar to Figure 3.24.

We can see that with four customers we are losing money. If we wish to examine them further, we could subset these four customers to another JMP table, as we did earlier in this chapter.

**Figure 3.24 Bar Chart Employing Further Elements of the Data Filter**

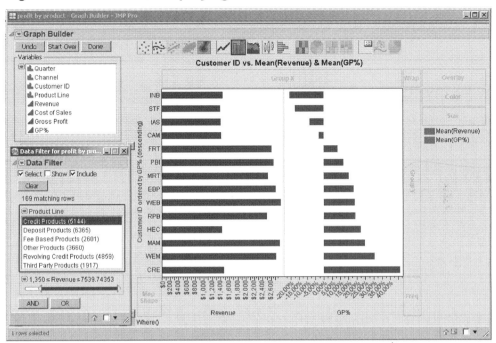

With the above examples, we have illustrated that by using several of the JMP visualization tools, and the additional layer of the data filter, we can quickly explore and develop a better understanding of a large amount of multivariate data.

---

[1] Thanks to Chuck Pirrello of SAS for providing the data set.
[2] Thanks to Chuck Pirrello of SAS for providing the data set.

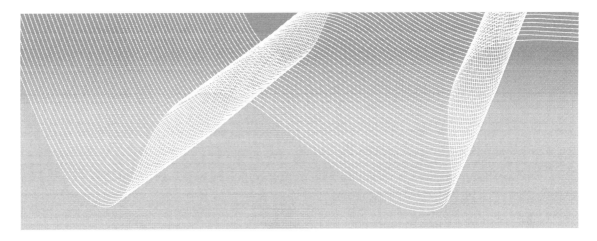

# Chapter 4

---

## Regression and ANOVA Review

Simple and multiple regression and ANOVA, as shown in our multivariate analysis framework in Figure 4.1, are two dependence techniques that are most likely covered in an introductory statistics course. In this chapter, we will review these two techniques, illustrate the process of performing these analyses in JMP, and explain how to interpret the results.

# Regression

One of the most popular applied statistical techniques is regression. Regression analysis typically has one of two main purposes. Either it is being used to understand the cause and effect relationship between one dependent variable and one or more independent variables. For example, how does the amount of advertising effect sales. Or regression is applied for prediction—in particular, for the purpose of forecasting a dependent variable based on one or more independent variables. An example would be using trend and seasonal components to forecast sales. Regression analyses can handle linear or nonlinear relationships, although the linear models are mostly emphasized.

## Simple Regression

Let's look at the data set **salesperfdata.jmp**. This data is from Cravens et al. (1972) and consists of the sales from 25 territories and seven corresponding variables:

- Salesperson's experience (Time)
- Market potential (MktPoten)
- Advertising expense (Adver)
- Market Share (MktShare)
- Change in Market Share (Change)
- Number of accounts (Accts)
- Workload per account (WkLoad)

## Figure 4.1  A Framework to Multivariate Analysis

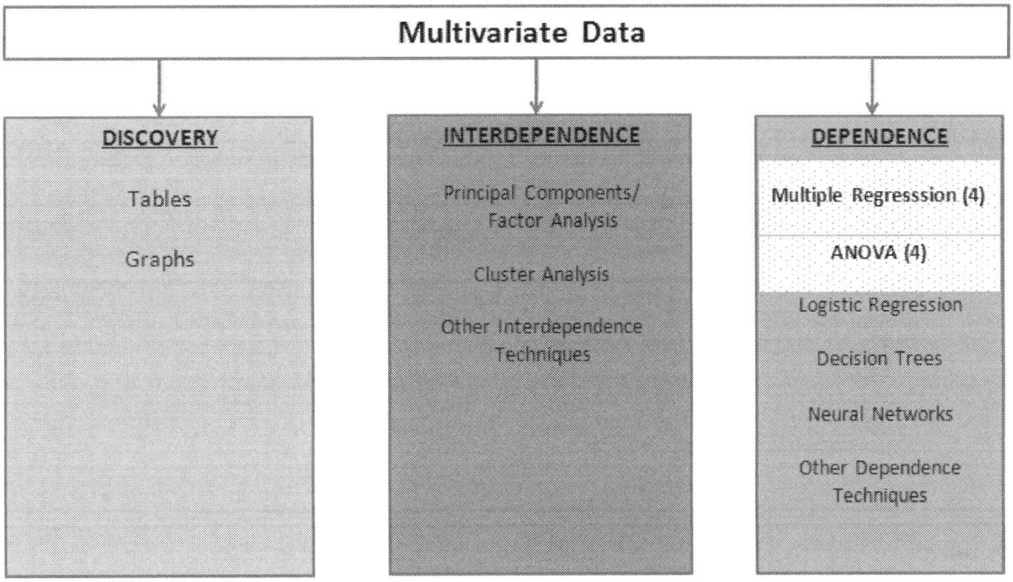

First, we will do only a simple linear regression, which is defined as one dependent variable Y and only one independent variable X. In particular, let's look at how advertising is related to sales:

> Open the salesperfdata.jmp in JMP. Select **Analyze→Fit Y by X**. In the Fit Y by X dialog box, click **Sales** and click Y, **Response**. Click **Adver** and click **X, Factor**. Click **OK**.

A scatterplot of sales versus advertising, similar to Figure 4.2 (without the red line), will appear. There appears to be a positive relationship, which means that as advertising increases, sales increase. To further examine and evaluate this relationship: Click the red triangle and click **Fit Line**. Fit Line causes the creation of the tables that contain the regression results. A simple linear regression is performed. The output will look like the bottom half of Figure 4.2. The regression equation is: Sales = 2106.09 + 0.2911*Adver. The y-intercept (black) and Adver (red) coefficient values are identified in Figure 4.2 under Parameter Estimates.

The F-test and *t*-test, in simple linear regression, are equivalent in that they both test whether the independent variable (**Adver**) is significantly related to the dependent variable (**Sales**). As indicated in Figure 4.2, both p-values are equal to 0.0017 and are significant. (To assist you in identifying significant relationships, JMP puts an asterisk next to a p-value that is significant with $\alpha = 0.05$.) This model's goodness-of-fit is measured by the coefficient of determination ($R^2$ or RSquare) and the standard error ($s_e$ or Root Mean Square Error), which in this case is equal to 35.5% and 1076.87, respectively.

**Figure 4.2  Scatterplot with Corresponding Simple Linear Regression**

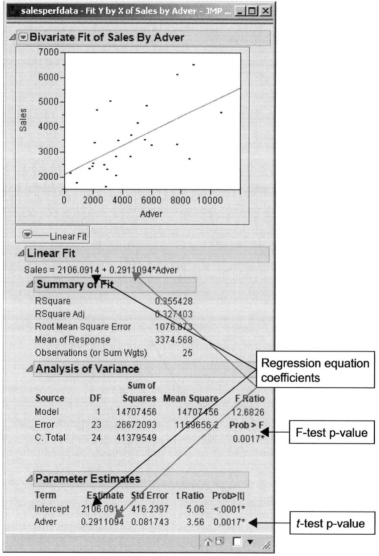

# Multiple Regression

We will next examine the multiple linear regression relationship between sales and the seven predictors (Time, MktPoten, Adver, MktShare, Change, Accts, and WkLoad), which is equivalent to those used by Cravens et al. (1972). Before performing the multiple linear regression, let's first examine the correlations and scatterplot matrix:

Select **Analyze→Multivariate Methods→Multivariate**. In the Multivariate and Correlations dialog box, in the Select Columns box, click **Sales**, hold down the shift key, and click **WkLoad.** Click **Y,Columns**, and all the variables will appear in the white box to the right of **Y, Columns**, as in Figure 4.3. Click **OK**.

## Figure 4.3 The Multivariate and Correlations Dialog Box

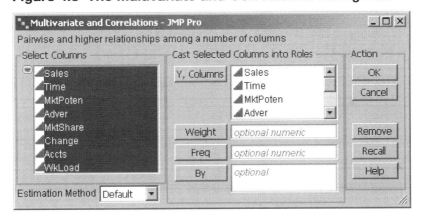

As shown in Figure 4.4, the correlation matrix and corresponding scatterplot matrix will be generated. Examine the first row of the Scatterplot Matrix. There appear to be several strong correlations, especially with the dependent variable Sales. All the variables, except for Wkload, have a strong positive relationship (or positive correlation) with Sales. The oval shape in each scatterplot is the corresponding bivariate normal density ellipse of the two variables. If the two variables are bivariate normally distributed, then, about 95% of the points would be within the ellipse. If the ellipse is rather wide (round) and does not follow either of the diagonals, then the two variables do not have a strong correlation. For example, observe the ellipse in the scatterplot for Sales and Wkload and notice that the correlation is -0.1172. The more significant the correlation, the more narrow the ellipse, and the more it follows along one of the diagonals (for example Sales and Accts).

To perform the multiple regression:

Select **Analyze→Fit Model**. In the Fit Model dialog box, as shown in Figure 4.5, click **Sales** and click **Y.** Next, click **Time,** hold down the shift key, and

click **WkLoad**. Now all the independent variables from Time to WkLoad are
highlighted. Click **Add** and now these independent variables are listed. Notice
the box to the right of Personality. It should say Standard Least Squares. Click
the drop-down arrow to the right and observe the different options, which
include stepwise regression (which we will discuss later in the chapter). Keep it
at Standard Least Squares and click **Run**.

**Figure 4.4  Correlations and Scatterplot Matrix for the salesperfdata.jmp
data**

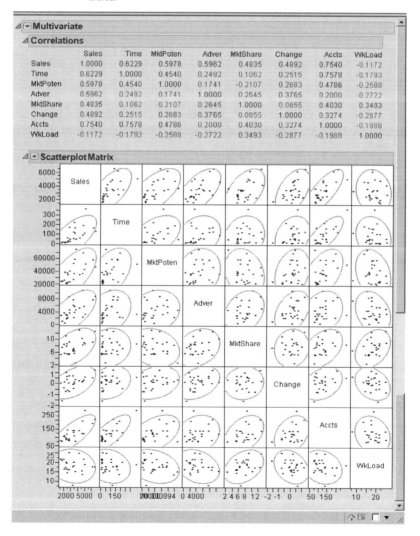

## Figure 4.5 The Fit Model Dialog Box

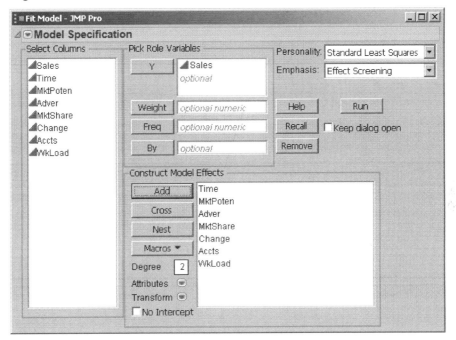

As shown in Figure 4.6, the multiple linear regression equation is:

Sales = -1485.88 + 1.97*Time + 0.04*MktPoten + 0.15*Adver + 198.31
*MktShare + 295.87*Change + 5.61* Accts +19.90*WkLoad.

Each independent variable regression coefficient represents an estimate of the change in the dependent variable that, in turn, corresponds to a unit increase in that independent variable while all the other independent variables are held constant. For example, the coefficient for WkLoad is 19.9. If WkLoad is increased by 1 and the other independent variables do not change, then Sales would correspondingly increase by 19.9.

The relative importance of each independent variable is measured by its standardized beta coefficient (which is a unitless value). The larger the absolute value of the standardized beta coefficient, the more important the variable. To obtain the standardized beta values, move the cursor somewhere over the Parameter Estimates Table and right-click. A list of options will appear, including **Columns**. Select **Columns→Std Beta**. Subsequently, the Std Beta column is added to the Parameter Estimates Table, as shown in Figure 4.7. With our sales performance data set, we can see that the variables MktPoten, Adver, and MktShare are the most important variables.

## Figure 4.6  Multiple Linear Regression Output

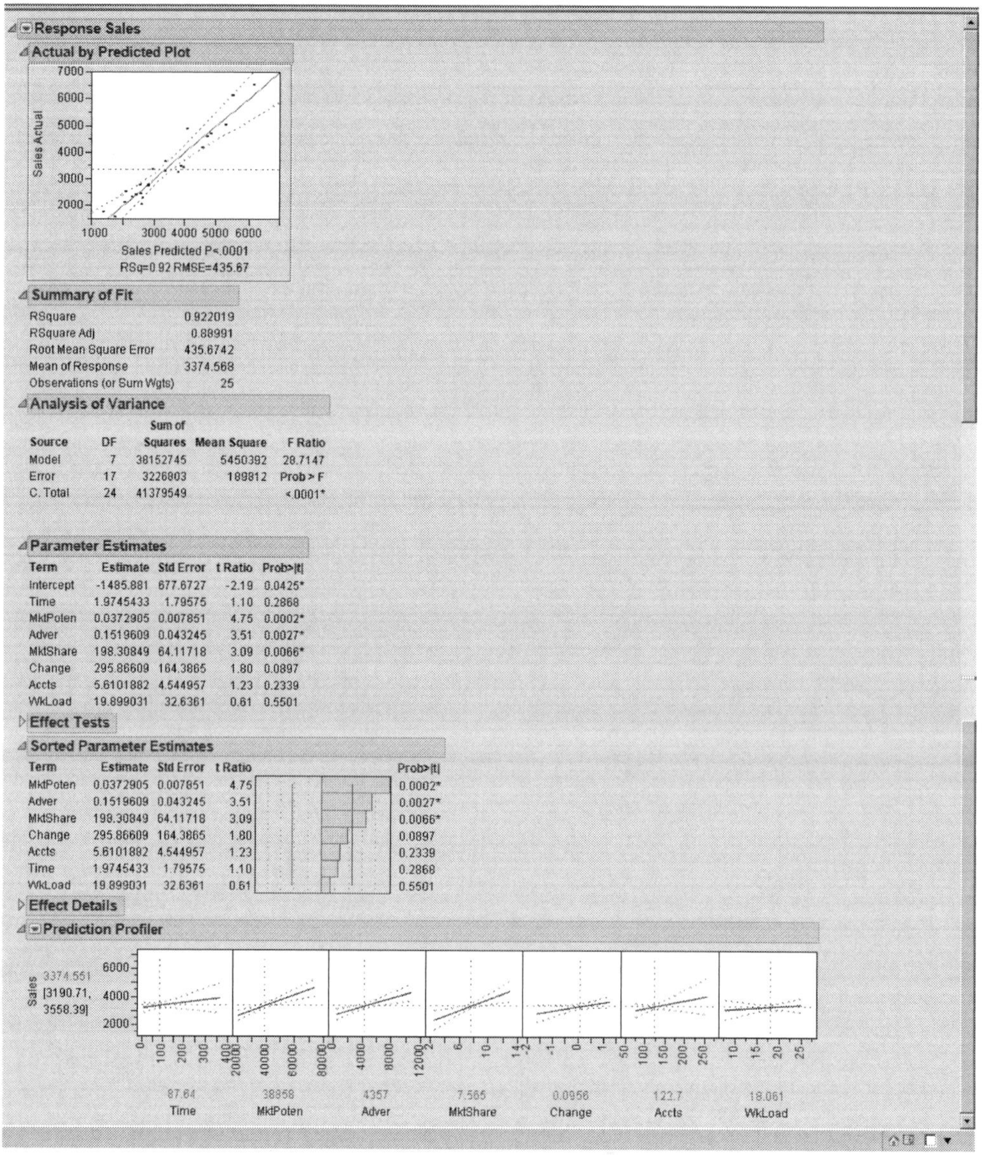

## Figure 4.7  Parameter Estimates Table with Standardized Betas and VIFs

| Term | Estimate | Std Error | t Ratio | Prob>\|t\| | Std Beta | VIF |
|------|----------|-----------|---------|---------|----------|-----|
| Intercept | -1485.881 | 677.6727 | -2.19 | 0.0425* | 0 | |
| Time | 1.9745433 | 1.79575 | 1.10 | 0.2868 | 0.130522 | 3.071766 |
| MktPoten | 0.0372905 | 0.007851 | 4.75 | 0.0002* | 0.446285 | 1.9246054 |
| Adver | 0.1519609 | 0.043245 | 3.51 | 0.0027* | 0.311209 | 1.709946 |
| MktShare | 198.30849 | 64.11718 | 3.09 | 0.0066* | 0.371494 | 3.1450824 |
| Change | 295.86609 | 164.3865 | 1.80 | 0.0897 | 0.139937 | 1.317864 |
| Accts | 5.6101882 | 4.544957 | 1.23 | 0.2339 | 0.194736 | 5.4257278 |
| WkLoad | 19.899031 | 32.6361 | 0.61 | 0.5501 | 0.055618 | 1.8139819 |

*Parameter Estimates*

The steps to evaluate a multiple regression model are listed in Figure 4.8. First, you should perform an F-test. The F-test, in multiple regression, is known as an overall test.

## Figure 4.8  Steps to Evaluating a Regression Model

### Process to Evaluating Statistical Significance of a Regression Model

<u>Good/Bad</u>
1. F-test
2. T-test for each independent variable
3. Look at residual plot
    (if time series data, D-W test)
4. Multicollinearity (VIF)

<u>How good of a fit</u>
1. Adjusted $R^2$
2. RMSE (or $s_e$)

The hypotheses for the F-test are:

$$H_0: \ \beta_1 = \beta_2 = \ldots = \beta_k = 0$$
$$H_1: \ \text{not all equal to } 0$$

; where k is the number of independent variables.

If we fail to reject the F-test, then the overall model is not statistically significantly. The model shows no linear significant relationships. We need to start over, and we should not look at the numbers in the Parameter Estimates table. On the other hand, if we reject $H_0$ of the F-test, then we can conclude that one or more of the independent variables are linearly related to the dependent variable. So we want to reject $H_0$ of the F-test.

In Figure 4.6, for our sales performance data set, we can see that the p-value for the F-test is <0.0001. So we reject the F-test and can conclude that one or more of the independent variables (i.e., one or more of variables Time, Mktpoten, Adver, Mktshare, Change, Accts, and Wkload) is significantly related to Sales (linearly).

The second step is to evaluate the t-test for each independent variable. Even though the hypotheses are:

$$H_0 : \quad \beta_k = 0$$
$$H_1 : \quad \beta_k \neq 0 \quad k = 1, 2, 3, \dots K$$

we are not testing whether independent variable k, $x_k$, is significantly related to the dependent variable. What we are testing is whether $x_k$ is significantly related to the dependent variable above and beyond all the other independent variables that are currently in the model. That is, in the regression equation, we have the term $\beta_k x_k$. If $\beta_k = 0$, then $x_k$ has no effect on Y. Again, we want to reject $H_0$. We can see in Figures 4.6 and 4.7 that the independent variables MktPoten, Adver, and MktShare each reject $H_0$ and are significantly related to Sales above and beyond the other independent variables.

One of the statistical assumptions of regression is that the residuals or errors (the difference between the actual value and the predicted value) should be random. That is, the errors do not follow any pattern. To visually examine the residuals, you can produce a plot of the residuals versus the predicted values by clicking the red triangle and selecting Row Diagnostics→Plot Residual by Predicted. A graph similar to Figure 4.9 will appear. Examine the plot for any patterns—oscillating too often or increasing or decreasing in values or for outliers. The data in Figure 4.9 appears to be random.

**Figure 4.9 Residual Plot of Predicted versus Residual**

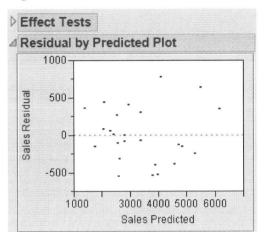

If the observations were taken in some time sequence, called a time series (not applicable to our sales performance data set), the Durbin-Watson test should be performed (click the red triangle next to the **Response Sales** heading; select **Row Diagnostics**→**Durbin-Watson Test**). To display the associated p-value, click the red triangle under the Durbin-Watson test and click **Significant p-value**. The Durbin-Watson test examines the correlation between consecutive observations. (In this case, you want high p-values; high p-values of the Durbin-Watson test indicate that there is no problem with first order autocorrelation.)

An important problem to address in the application of multiple regression is multicollinearity or collinearity of the independent variables (Xs) as in step 4 in Figure 4.8. Multicollinearity occurs when the two or more independent variables explain the same variability of Y. Multicollinearity does not violate any of the statistical assumptions of regression. However, significant multicollinearity is likely to make it difficult to interpret the meaning of the regression coefficients of the independent variables.

One method of measuring multicollinearity is the variance inflation factor (VIF). Each independent variable k has its own $VIF_k$. By definition, it must be greater than or equal to 1. The closer the $VIF_k$ is to 1, the smaller the relationship between the kth independent variable and the remaining Xs. Although, there are no definitive rules for identifying a large $VIF_k$, the basic guidelines (Marquardt, 1980) (Snee, 1973) for identifying whether significant multicollinearity exists are:

$1 \leq VIF_k \leq 5$   no significant multicollinearity

$5 < VIF_k \leq 10$   be concerned that some multicollinearity may exist

$VIF_k > 10$   significant multicollinearity

Note: Exceptions to these guidelines are when you may have transformed the variables (such as nonlinear transformation). To display the $VIF_k$ values for the independent variables (similar to what we illustrated to display the standardized beta values), move the cursor somewhere over the Parameter Estimates Table and right-click. Select **Columns→VIF**. As shown before in Figure 4.7, a column of VIF values for each independent variable is added to the Parameter Estimate Table. The variable Accts has a VIF greater than 5, and we should be concerned about that variable.

There are two opposing point of views as to whether to include those non-significant independent variables (i.e., those variables for which we failed to reject $H_0$ for the t-test) or the high $VIF_k$ variables. One perspective is to include those non-significant/high $VIF_k$ variables because they explain some, although not much, of the variation in the dependent variable, and it does improve your understanding of these relationships. Taking this approach to the extremes, you can keep on adding independent variables to the model so that you have almost the same number of independent variables as there are observations, which is somewhat unreasonable. However, this approach is sensible with a reasonable number of independent variables and if the major objective of the regression analysis is understanding the relationships of the independent variables to the dependent variable. Nevertheless, addressing the high $VIF_k$ values (especially $> 10$) may be of concern, particularly if one of the objectives of performing the regression is the interpretation of the regression coefficients.

The other point of view follows the principle of parsimony, which states that the smaller the number of variables in the model, the better. This viewpoint is especially true when the regression model is used for prediction/forecasting (having such variables in the model may actually increase the $s_e$). There are numerous approaches and several statistical variable selection techniques to achieve this goal of only significant independent variables in the model. Stepwise regression is one of the simplest approaches. Although it's not guaranteed to find the "best" model, it certainly will provide a model that is close to the 'best." To perform stepwise regression, we have two alternatives:

1. Click the red triangle next to the Response Sales heading and select **Model Dialog**. The Fit Model dialog box as in Figure 4.5 will appear.

2. Select **Analyze→Fit Model**. Click the **Recall** icon. The dialog box is repopulated and looks like Figure 4.5.

In either case, now, in the Fit Model dialog box, click the drop-down arrow to the right of Personality. Change Standard Least Squares to Stepwise. Click **Run**.

The Fit Stepwise dialog box will appear, similar to the top portion of Figure 4.10. In the drop-down box to the right of Stopping Rule, click the drop-down arrow, and change the default from Minimum BIC to P-Value Threshold. Two rows will appear, Prob to Enter and Prob to Leave. Change Direction to Mixed. (The Mixed option alternates the forward and backward steps.) Click **Go**.

Figure 4.10 displays the stepwise regression results. The stepwise model did not include the variables Accts and Wkload. The Adjusted $R^2$ values improved slightly from 0.890 in the full model to 0.893; and the $s_e$ (or Root Mean Square Error (RMSE)) actually improved by decreasing from 435.67 to 430.23. And nicely now, all the p-values for the t-tests are now significant (less than 0.05).

**Figure 4.10 Stepwise Regression for the Sales Performance Data Set**

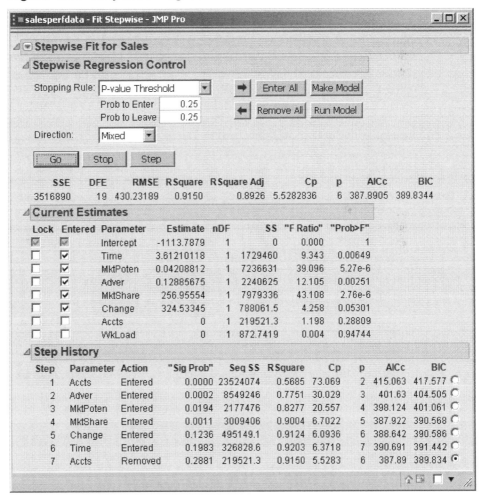

How good of a fit the regression model is measured by the Adjusted $R^2$ and the $s_e$ or RMSE, as listed in Figure 4.8. The Adjusted $R^2$ measures the percentage of the variability in the dependent variable that is explained by the set of independent variables and is adjusted for the number of independent variables (Xs) in the model. If the purpose of performing the regression is to understand the relationships of the independent variables with the dependent variable, the Adjusted $R^2$ value is a major assessment of the goodness of fit. What constitutes a good Adjusted $R^2$ value is subjective and depends on the situation. The higher the Adjusted $R^2$, the better the fit. On the other hand, if the regression model is for prediction/forecasting, the value of the $s_e$ is of more concern. A smaller $s_e$ generally means a smaller forecasting error. Similar to Adjusted $R^2$, a good $s_e$ value is subjective and depends on the situation.

Two additional approaches to consider in the model selection process that do consider both the model fit and the number of parameters in the model are the information criterion (AIC) approach developed by Akaike and the Bayesian information criterion (BIC) developed by Schwarz. The two approaches have quite similar equations except that the BIC criterion imposes a greater penalty on the number of parameters than the AIC criterion. Nevertheless, in practice, the two criteria often produce identical results. To apply these criteria, in the Stepwise dialog box when you click the drop-down arrow for Stopping Rule there were several options listed, including **Minimum AICc** and **Minimum BIC**.

## Regression with Categorical Data

There are many situations in which a regression model that uses one or more categorical variable(s) as (an) independent variable(s) may be of interest. For example, let's say our dependent variable is sales. A few possible illustrations of using a categorical variable that could affect sales would be gender, region, store, or month. The goal of using these categorical variables would be to see if the categorical variable may or may not significantly affect the dependent variable sales. That is, does the amount of sales differ significantly by gender, by region, by store, or by month? Or do sales vary significantly in combinations of several of these categorical variables?

Measuring ratios and distances are not appropriate with categorical data. That is, distance cannot be measured between male and female. So in order to use a categorical variable in a regression model, we must transform the categorical variable(s) into (a) continuous variable(s) or (an) integer (binary) variable(s). The resulting variables from this transformation are called indicator or dummy variables. How many indicator/dummy variables are required for a particular independent categorical variable X is equal to c-1, where c is the number of categories (or levels) of the categorical X variable. For example, gender requires one dummy variable, and month requires 11 dummy variables. In many cases, the dummy variables are coded 0 or 1. However, in JMP this is not the case. If a categorical variable has two categories (or levels) such as gender, then a single dummy variable is used with values of +1 and -1 (with +1 assigned to the alphabetically first category). If a categorical variable has more than two categories (or levels), the dummy variables are assigned values +1, 0 and -1.

Let's look at a data set, in particular AgeProcessWords.jmp (Eysenck, 1974). Eysenck conducted a study to examine whether age and the amount remembered is related to how the information was initially processed. Fifty younger and fifty older people were randomly assigned to five learning process groups (Counting, Rhyming, Adjective, Imagery, and Intentional), where younger was defined as being younger than 54 years old. The first four learning process groups were instructed on ways to memorize the 27 items. For example, the rhyming group was instructed to read each word and think of a word that rhymed with it. These four groups were not told that they would later be asked to recall the words. On the other hand, the last group, the intentional group, was told that they would be asked to recall the words. The data set has three variables: Age, Process, and Words.

To run a simple regression using the categorical variable Age to see how it is related to the number of words memorized, the variable Words:

> Select **Analyze**→**Fit Model**. In the Fit Model dialog box, select **Words**→**Y**. Then select **Age**→**Add**. Click **Run**.

Figure 4.11 shows the regression results. We can see from the F-test and t-test that Age is significantly related to Words. The regression equation is: Words = 11.61 - 1.55*Age[Older].

As we mentioned earlier, with a categorical variable with two levels, JMP assigns +1 to the alphabetically first level and -1 to the other level. So in this case, Age[Older] is assigned +1 and Age[Younger] is assigned -1. So the regression equations for each Age group are:

Older Age group: Words = 11.61 -1.55*Age[Older] = 11.61 -1.55(1) = 10.06
Younger Age group: Words = 11.61 -1.55*Age[Older] = 11.61 -1.55(-1) = 13.16

Therefore, the Younger age group can remember more words.

To verify the coding (and if you wish to create a different coding in other situations), we can create a new variable called Age1 by:

> Select **Cols**→**New Columns**. The new column dialog box will appear. In the text box for Column name, where Column 4 appears, type Age1. Click the drop-down arrow for Column Properties, and click **Formula**. The Formula dialog box will appear. In the text box below on the right for Functions (Grouped), select **Conditional**→**If**. The If statement will appear in the formula area. JMP has placed the cursor on expr (red rectangle). Click **Age** from the Table Columns. In the text box below on the right for Functions (Grouped), click

**Comparison>a==b**. In the new red rectangle, type "Older" (include quotation marks), and then press the **Tab** key. Click on the top then clause, type 1, and then press the **Tab** key again. Double-click on the lower else clause and type −1. The dialog box should look like Figure 4.12. Click **OK** and then click **OK** again.

## Figure 4.11 Regression of Age on Number of Words Memorized

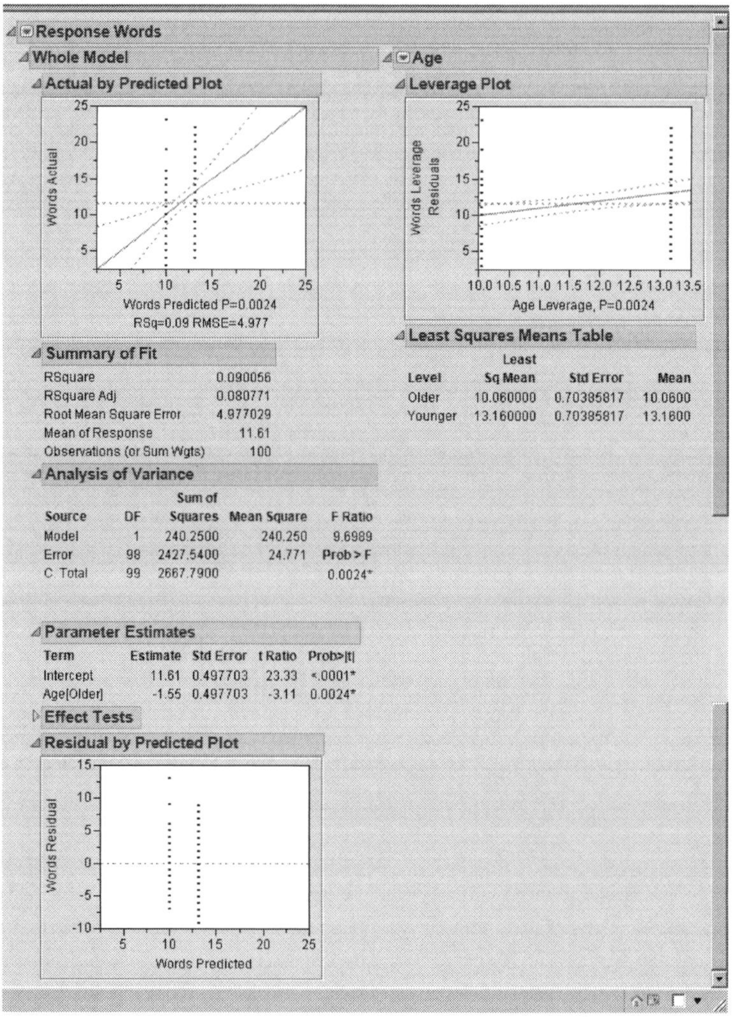

### Figure 4.12  Formula Dialog Box

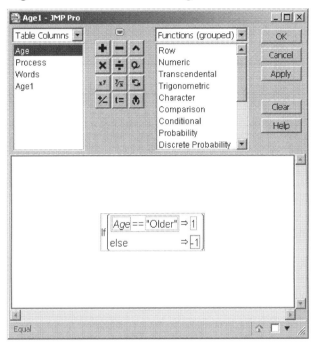

A new column **Age1** with +1s and −1s will be in the data table. Now, we run a regression with **Age1** as the independent variable. Figure 4.13 displays the results. Comparing the results in Figure 4.13 to the results in Figure 4.11, we see that they are exactly the same.

A regression that examines the relationship between **Words** and the categorical variable **Process** would use four dummy variables.

> Select **Analyze**→**Fit Model**. In the Fit Model dialog box, select **Process**→Y. Select **Age**→**Add**. Click **Run**.

## Figure 4.13 Regression of Age1 on Number of Words Memorized

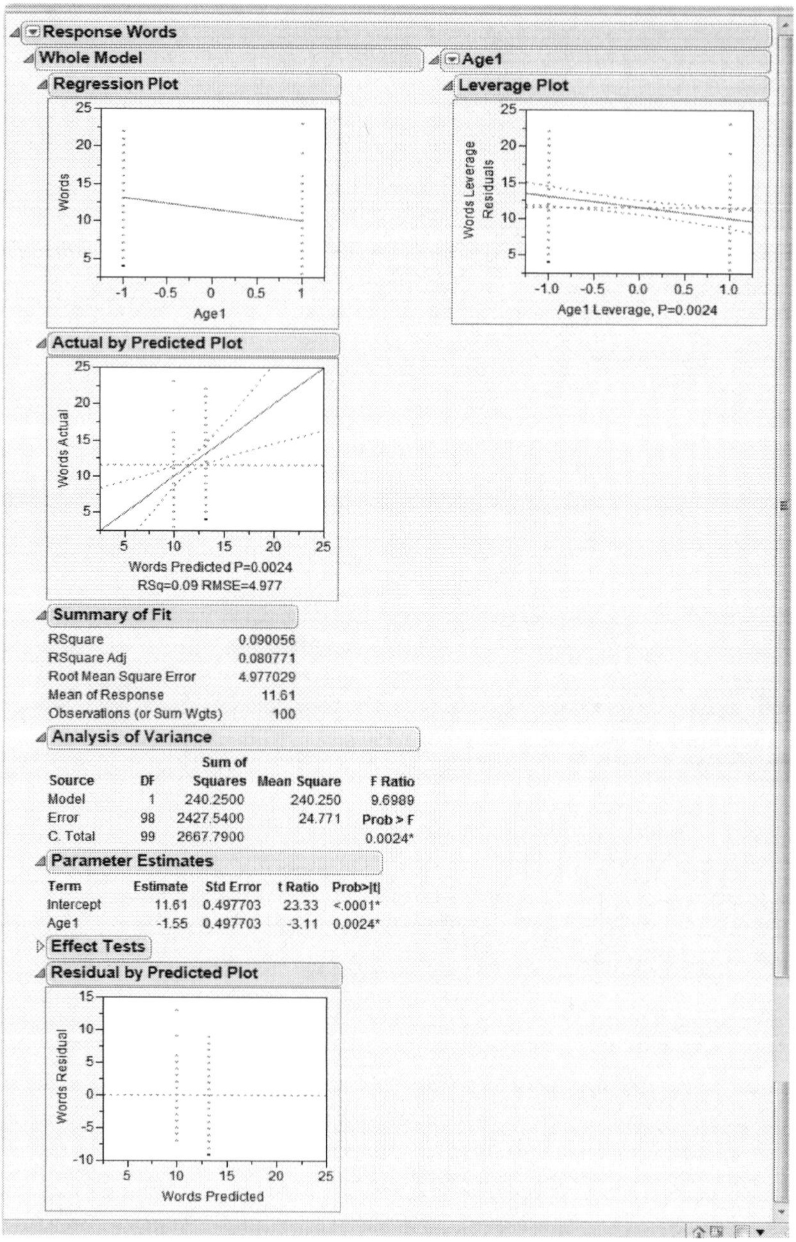

**Figure 4.14  Dummy Variable Coding for Process and the Predicted Number of Words Memorized**

| Row Process | Predicted # of Words | Process Dummy Variable | | | |
|---|---|---|---|---|---|
| | | [Adjective] | [Counting] | [Imagery] | [Intentional] |
| Adjective | 12.9 | 1 | 0 | 0 | 0 |
| Counting | 6.75 | 0 | 1 | 0 | 0 |
| Imagery | 15.5 | 0 | 0 | 1 | 0 |
| Intentional | 15.65 | 0 | 0 | 0 | 1 |
| Rhyming | 7.25 | -1 | -1 | -1 | -1 |

**Figure 4.15  Regression of Process on Number of Words Memorized**

◢ **Summary of Fit**

| | |
|---|---|
| RSquare | 0.567863 |
| RSquare Adj | 0.549668 |
| Root Mean Square Error | 3.48357 |
| Mean of Response | 11.61 |
| Observations (or Sum Wgts) | 100 |

◢ **Analysis of Variance**

| Source | DF | Sum of Squares | Mean Square | F Ratio |
|---|---|---|---|---|
| Model | 4 | 1514.9400 | 378.735 | 31.2095 |
| Error | 95 | 1152.8500 | 12.135 | Prob > F |
| C. Total | 99 | 2667.7900 | | <.0001* |

◢ **Parameter Estimates**

| Term | Estimate | Std Error | t Ratio | Prob>|t| |
|---|---|---|---|---|
| Intercept | 11.61 | 0.348357 | 33.33 | <.0001* |
| Process[Adjective] | 1.29 | 0.696714 | 1.85 | 0.0672 |
| Process[Counting] | -4.86 | 0.696714 | -6.98 | <.0001* |
| Process[Imagery] | 3.89 | 0.696714 | 5.58 | <.0001* |
| Process[Intentional] | 4.04 | 0.696714 | 5.80 | <.0001* |

◢ **Effect Tests**

| Source | Nparm | DF | Sum of Squares | F Ratio | Prob > F |
|---|---|---|---|---|---|
| Process | 4 | 4 | 1514.9400 | 31.2095 | <.0001* |

The coding used for these four **Process** dummy variables is shown in Figure 4.14. Figure 4.15 displays the regression results. The regression equation is:

**Words** = 11.61 +1.29\*Adjective – 4.86\*Counting + 3.89\*Imagery + 4.04\*Intentional

Given the dummy variable coding used by JMP, as shown in Figure 4.14, the mean of the level displayed in the brackets is equal to the difference between its coefficient value and the mean across all levels (that is, the overall mean or the y-intercept). So the t-tests here are testing if the level shown is different from the mean across all levels. If we haved used the dummy variable coding and substituted into the equation, the predicted number of Words that are memorized by process are as listed in Figure 4.14. Intentional has the highest words memorized at 15.65, and Counting is the lowest at 6.75.

# ANOVA

Analysis of variance (more commonly called ANOVA), as shown in Figure 4.1, is a dependence multivariate technique. There are several variations of ANOVA, such as one-factor (or one-way) ANOVA, two-factor (or two-way) ANOVA, and so on, and also repeated measures ANOVA. (In JMP, repeated measures ANOVA is found under **Analysis→Modeling→Categorical.**) The factors are the independent variables, each of which must be a categorical variable. The dependent variable is one continuous variable. We will provide only a brief introduction/review to ANOVA.

Let's assume one independent categorical variable X has k levels. One of the main objectives of ANOVA is to compare the means of two or more populations, i.e.:

$$H_0 = \mu_1 = \mu_2 \ldots = \mu_k$$
$$H_1 = \text{at least one mean is different from the other means}$$

The average for each level is $\mu_k$. If $H_0$ is true, we would expect all the sample means to be close to each other and relatively close to the grand mean. If $H_1$ is true, then at least one of the sample means would be significantly different. We measure the variability between means with the sum of squares between groups (SSBG). On the other hand, large variability within the sample weakens the capacity of the sample means to measure their corresponding population means. This within-sample variability is measured by the sum of squares within groups (or error) (SSE). As with regression, the decomposition of the Total sum of squares (TSS) is equal to:

Total Sum of Squares = Sum of Squares Between Groups + Sum of Squares Error or
TSS = SSBG + SSE.

(In JMP: TSS, SSBG, and SSE are identified as C.Total, Model SS, and Error SS, respectively.) To test this hypothesis, we do an F-test (the same as what is done in regression).

If $H_0$ of the F-test is rejected, which implies that one or more of the population means are significantly different, we then proceed to the second part of an ANOVA study and identify which factor level means are significantly different. If we have two populations (that is, two levels of the factor), we would not need ANOVA and we could perform a two-sample hypothesis test. On the other hand, when we have more than two populations (or groups) we could take the two-sample hypothesis test approach and compare every pair of groups. This approach has problems that are caused by multiple comparisons and is not advisable, as we will explain. If there are k levels in the X variable, the number of possible pairs of levels to compare is k(k-1)/2. For example, if we had three populations (call them A, B, and C), there would be 3 pairs to compare: A to B, A to C, and B to C. If we use a level of significance $\alpha = 0.05$, there is a 14.3% $(1-(.95)^3)$ chance that we will detect a difference in one of these pairs, even when there really is no difference in the populations; the probability of Type I error is 14.3%. If there are 4 groups, the likelihood of this error increases to 26.5%. This inflated error rate for the group of comparisons, not the individual comparisons, is not a good situation. However, ANOVA provides us with several multiple comparison tests that maintain the Type I error at $\alpha$ or smaller.

An additional plus of ANOVA is, if we are examining the relationship of two or more factors, ANOVA is good at uncovering any significant interactions or relationships among these factors. We will discuss interactions briefly when we examine two-factor (two-way) ANOVA. For now, let's look at one-factor (one-way) ANOVA.

## One-way ANOVA

One-way ANOVA has one dependent variable and one-X factor. Let's return to the AgeProcessWords.jmp data set and first perform a one-factor ANOVA with Words and Age:

> Select **Analyze→Fit Y by X**. In the Fit Y by X dialog box, select **Words→Y, Response**. Select **Age→X, Factor**. Click **OK**. In the Fit Y by X output, click the red triangle and select the Means/Anova/Pooled t option.

Figure 4.16 displays the one-way ANOVA results. The plot of the data at the top of Figure 4.16 is displayed by each factor level (in this case, the age group Older and Younger). The horizontal line across the entire plot represents the overall mean. Each factor level has its own mean diamond. The horizontal line in the center of the diamond is the mean for that level. The upper and lower vertices of the diamond represent the upper and lower 95% confidence limit on the mean, respectively. Additionally, the horizontal width of the diamond is relative to that level's (group's) sample size; that is, the wider the diamond, the larger the sample size for that level relative to the other levels. In this case, because the level sample sizes are the same, the horizontal widths of all the diamonds are the same. As shown in Figure 4.16, the t-test and the ANOVA F-test (in this situation, since the factor **Age** has only two levels, a pooled t-test is performed) show that there is a significant difference in the average **Words** memorized by **Age** level (p = 0.0024). In particular, examining the **Means for Oneway ANOVA** table in Figure 4.16, we can see that the Younger age group memorizes more words than the Older group (average of 13.16 to 10.06). Lastly, compare the values in the ANOVA table in Figure 4.16 to Figures 4.11 and 4.13. They have exactly the same values in the ANOVA tables when we did the simple categorical regression using AGE or AGE1.

## Figure 4.16 One-way ANOVA of Age and Words

**Oneway Analysis of Words By Age**

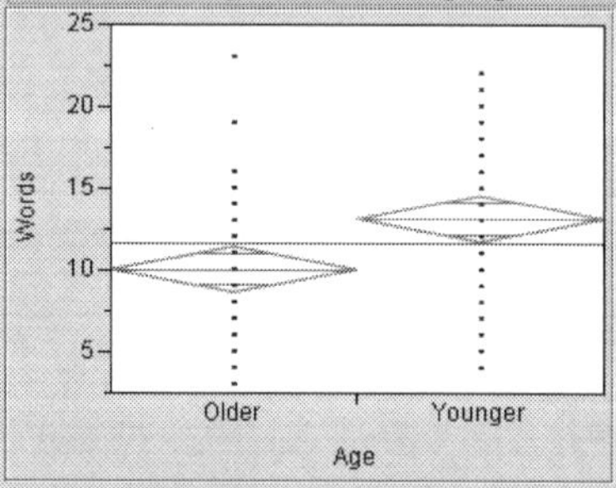

**Oneway Anova**

**Summary of Fit**

| | |
|---|---|
| Rsquare | 0.090056 |
| Adj Rsquare | 0.080771 |
| Root Mean Square Error | 4.977029 |
| Mean of Response | 11.61 |
| Observations (or Sum Wgts) | 100 |

**t Test**

Younger-Older
Assuming equal variances

| | | | |
|---|---|---|---|
| Difference | 3.10000 | t Ratio | 3.114308 |
| Std Err Dif | 0.99541 | DF | 98 |
| Upper CL Dif | 5.07535 | Prob > |t| | 0.0024* |
| Lower CL Dif | 1.12465 | Prob > t | 0.0012* |
| Confidence | 0.95 | Prob < t | 0.9988 |

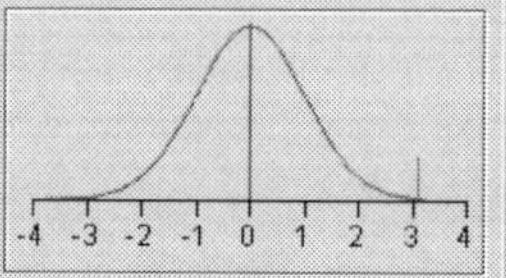

**Analysis of Variance**

| Source | DF | Sum of Squares | Mean Square | F Ratio | Prob > F |
|---|---|---|---|---|---|
| Age | 1 | 240.2500 | 240.250 | 9.6989 | 0.0024* |
| Error | 98 | 2427.5400 | 24.771 | | |
| C. Total | 99 | 2667.7900 | | | |

**Means for Oneway Anova**

| Level | Number | Mean | Std Error | Lower 95% | Upper 95% |
|---|---|---|---|---|---|
| Older | 50 | 10.0600 | 0.70386 | 8.663 | 11.457 |
| Younger | 50 | 13.1600 | 0.70386 | 11.763 | 14.557 |

Std Error uses a pooled estimate of error variance

# Testing Statistical Assumptions

ANOVA has three statistical assumptions to check and address. The first assumption is that the residuals should be independent. In most situations, ANOVA is rather robust in terms of this assumption. Furthermore, many times in practice this assumption is violated. So unless there is strong concern about the dependence of the residuals, this assumption does not have to be checked.

The second statistical assumption is that the variances for each level are equal. Violation of this assumption is of more concern because it could lead to erroneous p-values and hence incorrect statistical conclusions. We can test this assumption by clicking the red triangle and then clicking **Unequal Variances**. Added to the JMP ANOVA results is the Tests that the Variances are Equal report. Four tests are always provided. However, if, as it is in this case, there are only two groups tested, then an F-test for unequal variance is also performed; this gives us five tests, as shown in Figure 4.17. If we fail to reject $H_0$ (that is, we have a large p-value), we have insufficient evidence to say that the variances are not equal. So we proceed as though they are equal. On the other hand, if we reject $H_0$, the variances can be assumed to be unequal, and the ANOVA output cannot be trusted. For our problem, because for all the tests for unequal variances, the p-values are small, we need to look at the Welch's Test, located at the bottom of Figure 4.17. The p-value for the Welch ANOVA is 0.0025, which is also rather small. So we can assume that there is a significant difference in the population mean of Older and Younger.

The third statistical assumption is that the residuals should be normally distributed. The F-test is very robust if the residuals are non-normal. That is, if slight departures from normality are detected, they will have no real effect on the F statistic. A normal quantile plot can confirm whether the residuals are normally distributed or not. To produce a normal quantile plot:

> In the Fit Y by X output, click the red triangle and select **Save→Save residuals**. A new variable called Words centered by Age will appear in the data table. Select **Analyze→Distribution**. In the Distribution dialog box, in the Select Columns area, select **Words centered by Age→ Y, Response**. Click **OK**. The Distribution output will appear. Click on the middle red triangle (next to Words centered by Age), and click **Normal Quantile Plot**.

**Figure 4.17  Tests of Unequal Variances and Welch's Test for Age and Words**

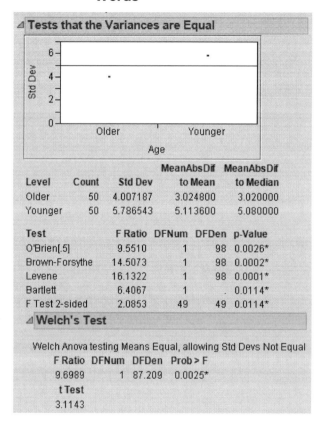

| Level | Count | Std Dev | MeanAbsDif to Mean | MeanAbsDif to Median |
|---|---|---|---|---|
| Older | 50 | 4.007187 | 3.024800 | 3.020000 |
| Younger | 50 | 5.786543 | 5.113600 | 5.080000 |

| Test | F Ratio | DFNum | DFDen | p-Value |
|---|---|---|---|---|
| O'Brien[.5] | 9.5510 | 1 | 98 | 0.0026* |
| Brown-Forsythe | 14.5073 | 1 | 98 | 0.0002* |
| Levene | 16.1322 | 1 | 98 | 0.0001* |
| Bartlett | 6.4067 | 1 | . | 0.0114* |
| F Test 2-sided | 2.0853 | 49 | 49 | 0.0114* |

◢ **Welch's Test**

Welch Anova testing Means Equal, allowing Std Devs Not Equal

| F Ratio | DFNum | DFDen | Prob > F |
|---|---|---|---|
| 9.6989 | 1 | 87.209 | 0.0025* |

| t Test |
|---|
| 3.1143 |

The distribution output and normal quantile plot will look like Figure 4.18. If all the residuals fall on or near the straight line or within the confidence bounds, the residuals should be considered normally distributed. All the points in Figure 4.18 are within the bounds, so we can assume that the residuals are normally distributed.

**Figure 4.18 Normal Quantile Plot of the Residuals**

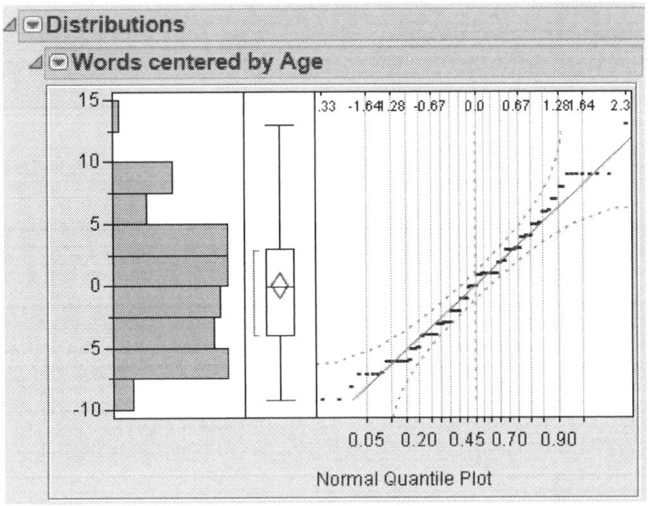

Let's now perform a one-factor ANOVA of Process and Words:

## Figure 4.19  One-way ANOVA of Process and Words

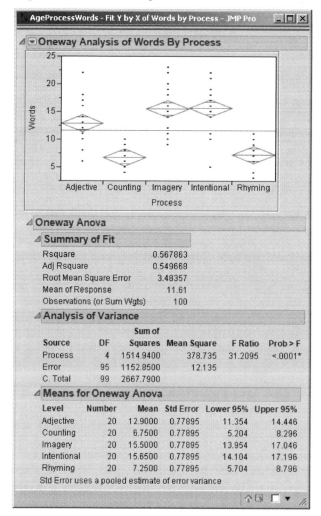

Select **Analyze→Fit Y by X**. In the Fit Y by X dialog box, select **Words→Y, Response**. Select **Process→X, Factor**. Click **OK**. Click the red triangle and click **Means/Anova**.

Figure 4.19 displays the one-way ANOVA results. The p-value for the F-test is <0.0001, so one or more of the Process means differ from each other. Also, notice that the ANOVA table in Figure 4.19 is the same as in the categorical regression output of Words and Process in Figure 4.15. The results of the tests for unequal variances are displayed in Figure 4.20.

**Figure 4.20 Tests of Unequal Variances and Welch's Test for Process and Words**

**△ Tests that the Variances are Equal**

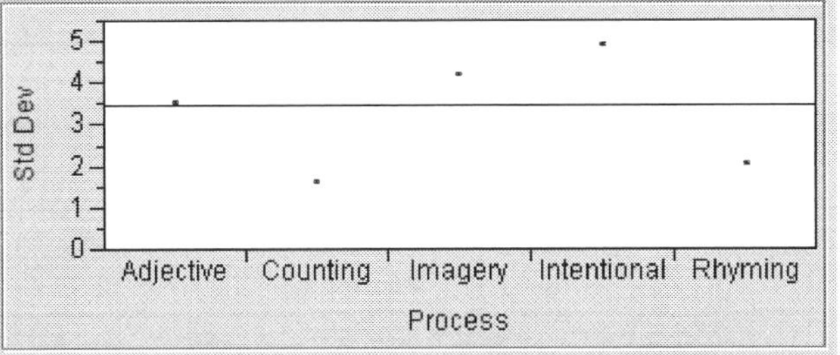

| Level | Count | Std Dev | MeanAbsDif to Mean | MeanAbsDif to Median |
|-------|-------|---------|--------------------|----------------------|
| Adjective | 20 | 3.537766 | 2.510000 | 2.500000 |
| Counting | 20 | 1.618154 | 1.275000 | 1.250000 |
| Imagery | 20 | 4.173853 | 3.450000 | 3.400000 |
| Intentional | 20 | 4.901933 | 4.050000 | 4.050000 |
| Rhyming | 20 | 2.022895 | 1.525000 | 1.450000 |

| Test | F Ratio | DFNum | DFDen | Prob > F |
|------|---------|-------|-------|----------|
| O'Brien[.5] | 5.0043 | 4 | 95 | 0.0011* |
| Brown-Forsythe | 6.9145 | 4 | 95 | <.0001* |
| Levene | 7.1390 | 4 | 95 | <.0001* |
| Bartlett | 7.1875 | 4 | . | <.0001* |

**△ Welch's Test**

Welch Anova testing Means Equal, allowing Std Devs Not Equal

| F Ratio | DFNum | DFDen | Prob > F |
|---------|-------|-------|----------|
| 38.4677 | 4 | 45.582 | <.0001* |

In general, the Levene test is more widely used and more comprehensive, so, we focus only on the Levene test. It appears that we should assume that the variances are unequal (for all the tests), so we should look at the Welch's Test in Figure 4.20. Because the p-value for the Welch's Test is small, we can reject the null hypothesis; the pairs of means are different from one another.

# Testing for Differences

The second part of the ANOVA study focuses on identifying which factor level means differ from each other. However, with the AgeProcessWords.jmp data set, because we did not satisfy the statistical assumption that the variances for each level are equal, we will not perform the second stage of the ANOVA study with that data set. (In general, it is not recommended to perform these second stage tests if the equal variances assumption is not satisfied. Nonetheless, if you do, JMP does provide some nonparametric tests.)

Consequently, let's use another data set, Analgesics.jmp. The Analgesics data table contains 33 observations and three variables, Gender, Drug (3 analgesics drugs: A, B, and C), and Pain (ranges from 2.13 and 16.64—in which the higher the value, the more painful). Figures 4.21 and 4.22 display the two one-way ANOVAs. Males appear to have significantly higher Pain, and there does seem to be significant differences in Pain by Drug. The p-value for the Levene test is 0.0587, so we fail to reject the null hypothesis of all equal variances and can comfortably move on to the second part of an ANOVA study. (On the other hand, in Figure 4.22 the diamonds under Oneway analysis of pain by Drug appear different, suggesting possible different variances. If we select the red triangle and click Unequal Variances, we can check the Welch's Test. Since the p-value for the Welch's Test is small, we can reject the null hypothesis; the pairs of means are different from one another.)

## Figure 4.21 One-way ANOVA of Gender and Pain

**Oneway Analysis of pain By gender**

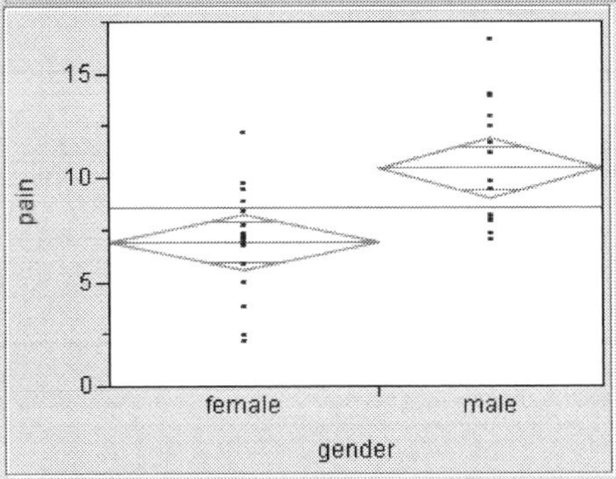

### Oneway Anova

#### Summary of Fit

| | |
|---|---|
| Rsquare | 0.307865 |
| Adj Rsquare | 0.285538 |
| Root Mean Square Error | 2.749419 |
| Mean of Response | 8.533515 |
| Observations (or Sum Wgts) | 33 |

#### t Test

male-female

Assuming equal variances

| | | | |
|---|---|---|---|
| Difference | 3.56929 | t Ratio | 3.713348 |
| Std Err Dif | 0.96120 | DF | 31 |
| Upper CL Dif | 5.52968 | Prob > |t| | 0.0008* |
| Lower CL Dif | 1.60890 | Prob > t | 0.0004* |
| Confidence | 0.95 | Prob < t | 0.9996 |

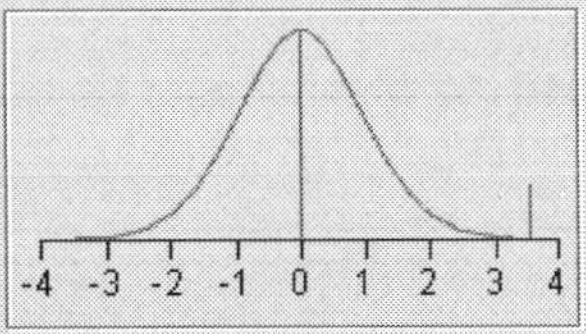

#### Analysis of Variance

| Source | DF | Sum of Squares | Mean Square | F Ratio | Prob > F |
|---|---|---|---|---|---|
| gender | 1 | 104.23492 | 104.235 | 13.7890 | 0.0008* |
| Error | 31 | 234.33844 | 7.559 | | |
| C. Total | 32 | 338.57335 | | | |

#### Means for Oneway Anova

| Level | Number | Mean | Std Error | Lower 95% | Upper 95% |
|---|---|---|---|---|---|
| female | 18 | 6.9111 | 0.64804 | 5.5894 | 8.233 |
| male | 15 | 10.4804 | 0.70990 | 9.0326 | 11.928 |

Std Error uses a pooled estimate of error variance

## Figure 4.22 One-way ANOVA of Drug and Pain

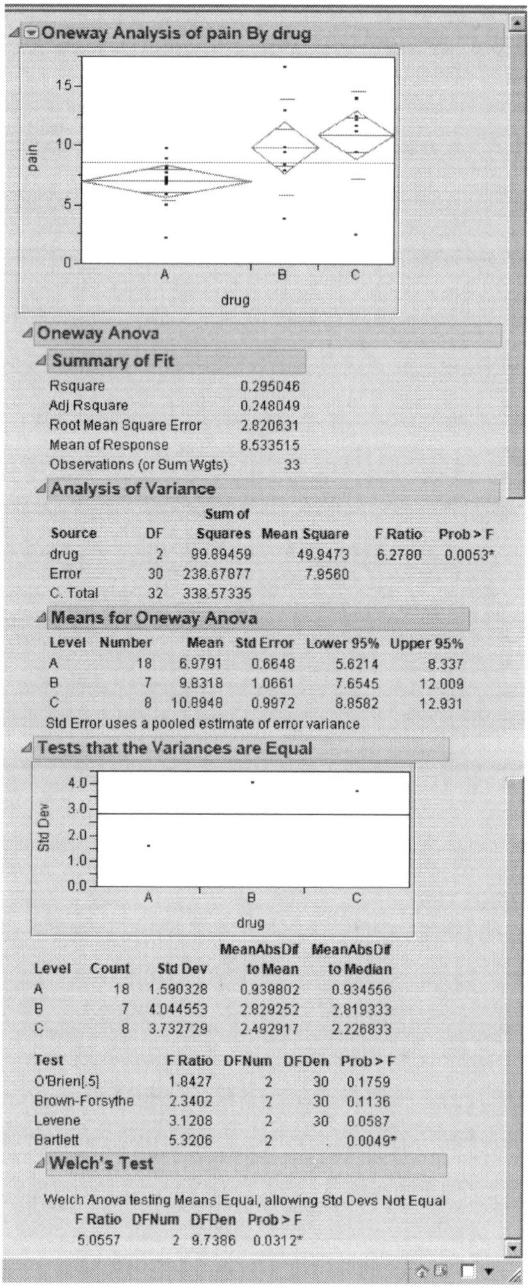
Oneway Analysis of pain By drug
15
10
5
0
pain
A
B
C
drug
Oneway Anova
Summary of Fit
Rsquare 0.295046
Adj Rsquare 0.248049
Root Mean Square Error 2.820631
Mean of Response 8.533515
Observations (or Sum Wgts) 33
Analysis of Variance
Source DF Sum of Squares Mean Square F Ratio Prob > F
drug 2 99.89459 49.9473 6.2780 0.0053*
Error 30 238.67877 7.9560
C. Total 32 338.57335
Means for Oneway Anova
Level Number Mean Std Error Lower 95% Upper 95%
A 18 6.9791 0.6648 5.6214 8.337
B 7 9.8318 1.0661 7.6545 12.009
C 8 10.8948 0.9972 8.8582 12.931
Std Error uses a pooled estimate of error variance
Tests that the Variances are Equal
4.0
3.0
2.0
1.0
0.0
Std Dev
A
B
C
drug
Level Count Std Dev MeanAbsDif to Mean MeanAbsDif to Median
A 18 1.590328 0.939802 0.934556
B 7 4.044553 2.829252 2.819333
C 8 3.732729 2.492917 2.226833
Test F Ratio DFNum DFDen Prob > F
O'Brien[.5] 1.8427 2 30 0.1759
Brown-Forsythe 2.3402 2 30 0.1136
Levene 3.1208 2 30 0.0587
Bartlett 5.3206 2 . 0.0049*
Welch's Test
Welch Anova testing Means Equal, allowing Std Devs Not Equal
F Ratio DFNum DFDen Prob > F
5.0557 2 9.7386 0.0312*

JMP provides several multiple comparison tests, which can be found by clicking the red triangle, clicking Compare Means, and then selecting the test. The first test, the Each Pair, Student's t-test, computes individual pairwise comparisons. As discussed earlier, the likelihood of a Type I error increases with the number of pairwise comparisons. So, unless the number of pairwise comparisons is small, we do not recommend using this test.

The second means comparison test is the All Pairs, Tukey HSD test. If the main objective is to check for any possible pairwise difference in the mean values, and there are several factor levels, the Tukey HSD (Honestly Significant Difference) or Tukey-Kramer HSD test is the most desired test. Figure 4.23 displays the results of the Tukey-Kramer HSD test. To identify mean differences, examine the Connecting Letters Report. Groups that do not share the same letter are significantly different from one another. The mean pain for Drug A is significantly different from the mean pain of Drug C.

## Figure 4.23 The Tukey-Kramer HSD Test

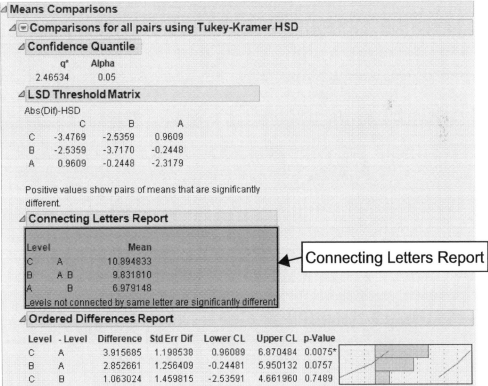

The third means comparison test is **With Best, Hsu MCB**. The Hsu's MCB (multiple comparison with best) is used to determine whether each factor level mean can be rejected or not as the "best" of all the other means, where "best" means either a maximum or minimum value. Figure 4.24 displays the results of Hsu's MCB test. The p-value report and the maximum and minimum LSD (Least Squares Differences) matrices can be used to identify significant differences. The p-value report identifies whether a factor level mean is significantly different from the maximum and from the minimum of all the other means. The **Drug A** is significantly different from the maximum, and **Drugs B** and **C** are significantly different from the minimum. The maximum LSD matrix compares the means of the groups against the unknown maximum and correspondingly the same for the minimum LSD matrix. Follow the directions below each matrix, looking at positive values for the maximum and negative values for the minimum.

### Figure 4.24  The Hsu's MCB Test

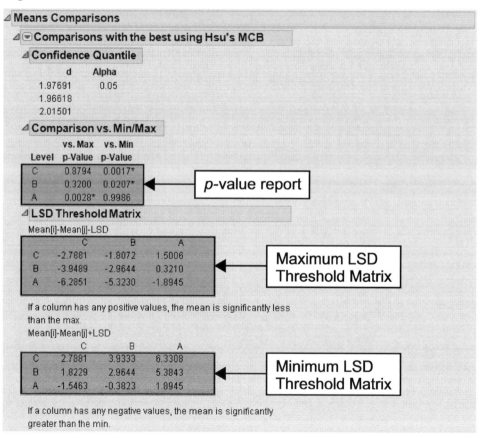

Examining the matrices in Figure 4.24, we can see that the Drug groups AC and AB are significantly less than the maximum and significantly greater than the minimum.

Differences with the Hsu's MCB test are less conservative than those found with the Tukey-Kramer test. Additionally, Hsu's MCB test should be used if there is a need to make specific inferences about the maximum or minimum values.

The last means comparison test provided, the with Control, Dunnett's, is applicable when you do not wish to make all pairwise comparisons, but rather only to compare one of the levels (the "control") with each other level. Thus, fewer pairwise comparisons are made. Because this is not a common situation, we will not discuss this test further.

In addition, one-way ANOVA can also be performed using the JMP Fit Model platform. Most of the output is identical to the Fit Y by X results, but, with a few differences. Let's perform a one-way ANOVA on Pain with Drug:

> Select **Analyze→Fit Model**. In the Fit Model dialog box, click **Pain→Y**. Select Drug→**Add**. Click **Run**. On the right side of the output, click the red triangle next to Drug and select **LSMeans Plot→LSMeans Tukey HSD**.

The output will look similar to Figure 4.25. Realize that these results are also equivalent to the regression results with these categorical variables that we discussed early in this chapter (see Figure 4.15). The LSMeans plot shows a plot of the factor means. The LSMeans Tukey HSD is similar to the All Pairs, Tukey HSD output from the Fit Y by X option. In Figure 4.25 and Figure 4.23, notice the same Connecting Letters Report. The Fit Model output does not have the LSD matrices but does provide us with a crosstab report. The Fit Model platform does not directly provide an equal variance test. But, you can visually evaluate the Residual by Predicted Plot in the lower left corner of Figure 4.25.

## Figure 4.25 One-way ANOVA Output Using Fit Model

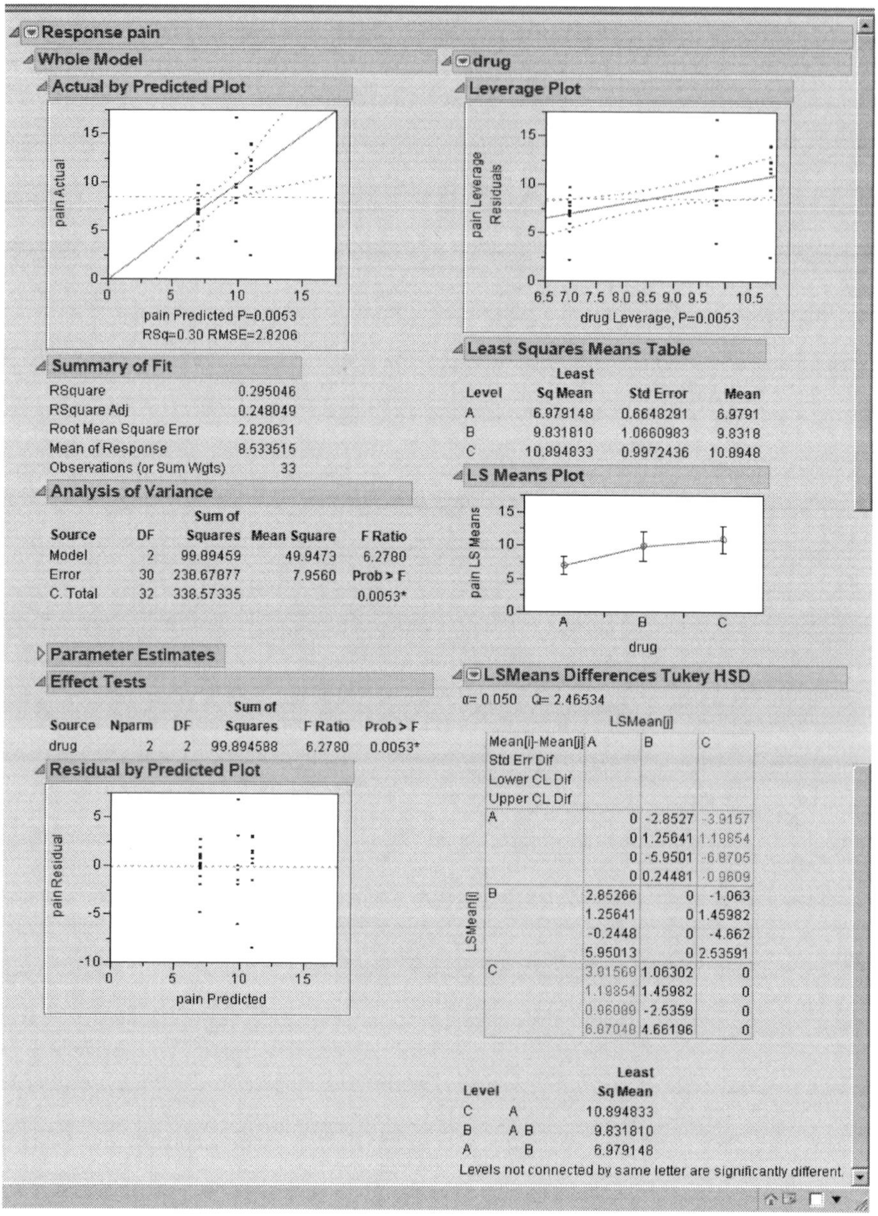

## Two-way ANOVA

Two-way ANOVA is an extension of the one-way ANOVA in which there is one continuous dependent variable, but, now we have two categorical independent variables. There are three basic two-way ANOVA designs: without replication; with equal replication, and; with unequal replication. The two two-way designs with replication not only allow us to address the influence of each factor's contribution to explaining the variation in the dependent variable, but, also allow us to address the interaction effects due to all the factor level combinations. We will only discuss the two-way ANOVA with equal replication.

Using the Analgesics.jmp, let's perform a two-way ANOVA of Pain with Gender and Drug as independent variables:

> Select **Analyze→Fit Model**. In the Fit Model dialog box, select **Pain→Y**. Click **Gender**, hold down the shift key, and click **Drug**. Then click **Add**. Again, click **Gender**, hold down the shift key, and click **Drug**. This time, click **Cross**. (Another approach: highlight Gender and Drug, click **Macros**, and click **Full Factorial**.) Click **Run**.

Figure 4.26 displays the two-way ANOVA results. The p-value for the F-test in the ANOVA table is 0.0006, so there are significant differences in the means. To find where the differences are, we examine the F-tests in the Effect Tests table. We can see that there are significant differences in the Age, Drug (which we found out already when we did the one-way ANOVAs). But we observe a moderately significant difference in the interaction means, p-value = 0.0916. To understand these differences, go back to the top of the output and click the red triangles for each factor and then:

> For Gender: Select **LSMeans Plot→LSMeans Student's t** (since we have only two levels).

> For Drug: Select **LSMeans Plot→LSMeans Tukey HSD**.

> For Gender*Drug: Select **LSMeans Plot→LSMeans Tukey HSD**.

The results for Gender and Drug displayed in Figure 4.26 are equivilant to what we observed earlier in the chapter when we discussed one-way ANOVA. Figure 4.27 also has the LSMeans Plot and the Connecting Letter Report for the interaction effect Gender*Drug. If there were no significant interaction, the lines in the LSMeans Plot would not cross and will be mostly parallel.

## Figure 4.26  Two-way ANOVA Output

### Response pain

#### Whole Model

##### Actual by Predicted Plot

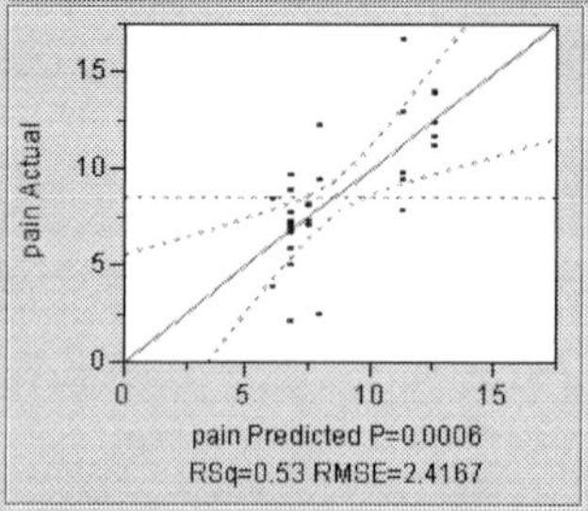

pain Predicted P=0.0006
RSq=0.53 RMSE=2.4167

##### Summary of Fit

| | |
|---|---|
| RSquare | 0.534258 |
| RSquare Adj | 0.448009 |
| Root Mean Square Error | 2.41667 |
| Mean of Response | 8.533515 |
| Observations (or Sum Wgts) | 33 |

##### Analysis of Variance

| Source | DF | Sum of Squares | Mean Square | F Ratio |
|---|---|---|---|---|
| Model | 5 | 180.88537 | 36.1771 | 6.1944 |
| Error | 27 | 157.68799 | 5.8403 | Prob > F |
| C. Total | 32 | 338.57335 | | 0.0006* |

##### Parameter Estimates

| Term | Estimate | Std Error | t Ratio | Prob>|t| |
|---|---|---|---|---|
| Intercept | 8.7207484 | 0.494984 | 17.62 | <.0001* |
| gender[female] | -1.759652 | 0.494984 | -3.55 | 0.0014* |
| drug[A] | -1.585892 | 0.616267 | -2.57 | 0.0159* |
| drug[B] | -0.009982 | 0.765306 | -0.01 | 0.9897 |
| gender[female]*drug[A] | 1.409308 | 0.616267 | 2.29 | 0.0303* |
| gender[female]*drug[B] | -0.856115 | 0.765306 | -1.12 | 0.2731 |

##### Effect Tests

| Source | Nparm | DF | Sum of Squares | F Ratio | Prob > F |
|---|---|---|---|---|---|
| gender | 1 | 1 | 73.808295 | 12.6378 | 0.0014* |
| drug | 2 | 2 | 51.059196 | 4.3713 | 0.0227* |
| gender*drug | 2 | 2 | 30.542763 | 2.6148 | 0.0916 |

##### Residual by Predicted Plot

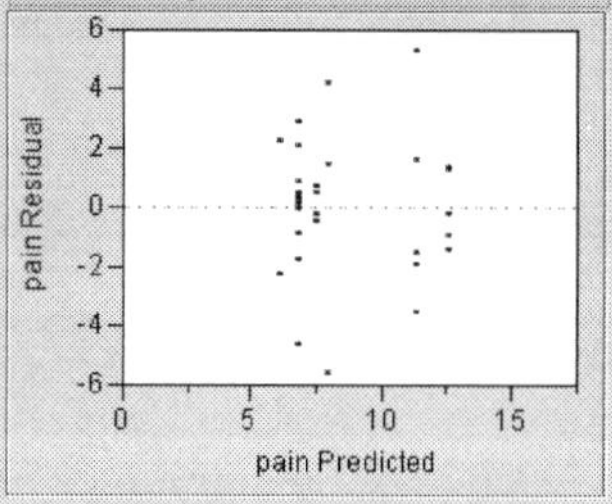

**Figure 4.27 Two-way ANOVA Output for Gender, Drug, and Gender*Drug**

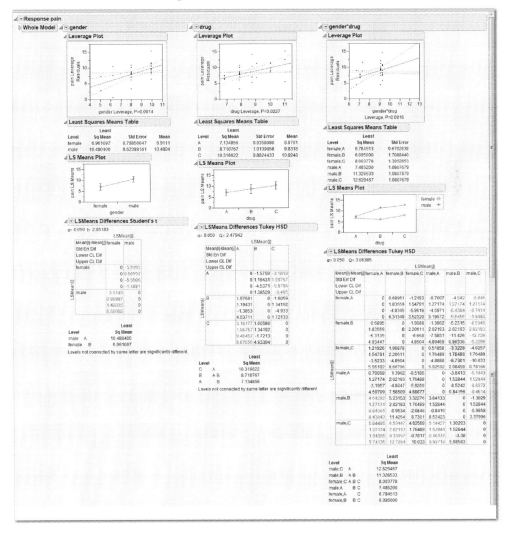

This is not the case here. The Connecting Letter Report identifies further the significant interactions. Overall, and clearly in the LS Means Plot, it appears that Females have lower Pain than Males. However, for Drug A, Females have higher than expected Pain.

To check for equal variances in two-way ANOVA, we need to create a new column that is the interaction, Gender*Drug, and then run a one-way ANOVA on that variable. To create a new column when both variables are characters:

Select **Cols**→**New Columns**. The new column dialog box will appear. In the text box for Column name, where Column 4 appears, type Gender*Drug. Click the drop-down arrow for Column Properties and click **Formula**. The Formula dialog box will appear. Click the drop-down arrow next to Functions (Grouped), and click **Functions (all)**. Scroll down until you see **Concat** and click it. The Concat statement will appear in the formula area with two rectangles. JMP has placed the cursor on the left rectangle (red rectangle). Click **Gender** from the Table Columns; click the right Concat rectangle and click **Drug** from the Table Columns. The dialog box should look like Figure 4.28. Click **OK** and then Click **OK** again.

**Figure 4.28  Formula Dialog Box**

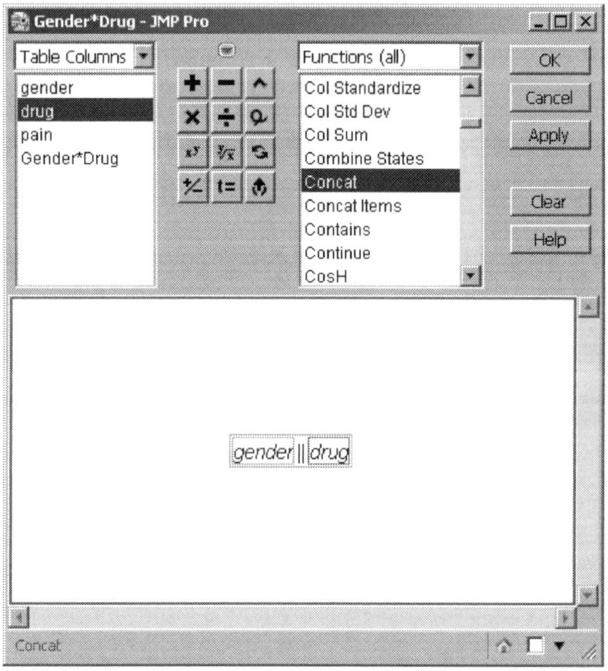

The results of running a one-way ANOVA of Gender*Drug and Pain is shown in Figure 4.29. The F-test rejects the null hypothesis. So, as we have seen before, there is significant interaction.

## Figure 4.29  One-way ANOVA with Pain and Interaction Gender*Drug

### Oneway Analysis of pain By Gender*Drug

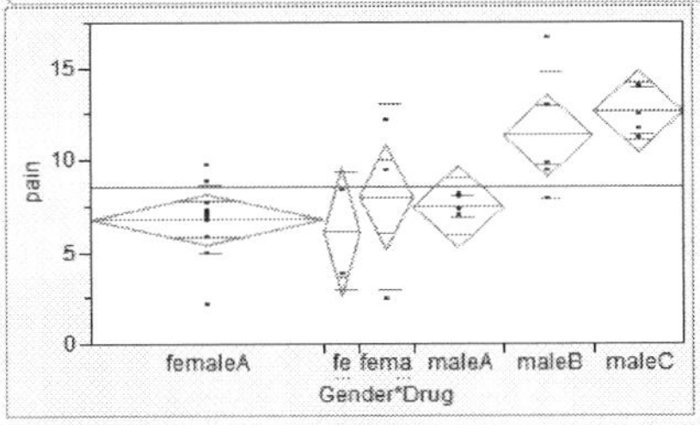

### Oneway Anova

#### Summary of Fit

| | |
|---|---|
| Rsquare | 0.534258 |
| Adj Rsquare | 0.448009 |
| Root Mean Square Error | 2.41667 |
| Mean of Response | 8.533515 |
| Observations (or Sum Wgts) | 33 |

#### Analysis of Variance

| Source | DF | Sum of Squares | Mean Square | F Ratio | Prob > F |
|---|---|---|---|---|---|
| Gender*Drug | 5 | 180.88537 | 36.1771 | 6.1944 | 0.0006* |
| Error | 27 | 157.68799 | 5.8403 | | |
| C. Total | 32 | 338.57335 | | | |

#### Means for Oneway Anova

| Level | Number | Mean | Std Error | Lower 95% | Upper 95% |
|---|---|---|---|---|---|
| femaleA | 13 | 6.7845 | 0.6703 | 5.409 | 8.160 |
| femaleB | 2 | 6.0950 | 1.7088 | 2.589 | 9.601 |
| femaleC | 3 | 8.0038 | 1.3953 | 5.141 | 10.867 |
| maleA | 5 | 7.4852 | 1.0808 | 5.268 | 9.703 |
| maleB | 5 | 11.3265 | 1.0808 | 9.109 | 13.544 |
| maleC | 5 | 12.6295 | 1.0808 | 10.412 | 14.847 |

Std Error uses a pooled estimate of error variance

#### Tests that the Variances are Equal

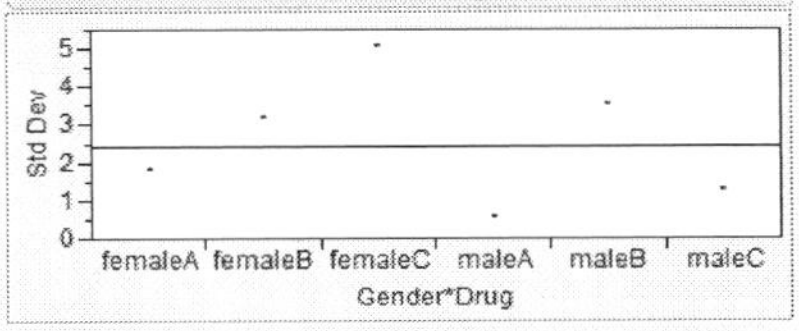

| Level | Count | Std Dev | MeanAbsDif to Mean | MeanAbsDif to Median |
|---|---|---|---|---|
| femaleA | 13 | 1.824411 | 1.149239 | 1.109436 |
| femaleB | 2 | 3.191880 | 2.257000 | 2.257000 |
| femaleC | 3 | 5.044267 | 3.732741 | 4.189556 |
| maleA | 5 | 0.565866 | 0.484907 | 0.485200 |
| maleB | 5 | 3.495135 | 2.765173 | 2.539733 |
| maleC | 5 | 1.277130 | 1.054293 | 1.158533 |

| Test | F Ratio | DFNum | DFDen | Prob > F |
|---|---|---|---|---|
| O'Brien[.5] | 2.8604 | 4 | 26 | 0.0434* |
| Brown-Forsythe | 2.9319 | 5 | 27 | 0.0306* |
| Levene | 3.9209 | 5 | 27 | 0.0084* |
| Bartlett | 2.9507 | 5 | | 0.0115* |

Warning: Small sample sizes. Use Caution.

#### Welch's Test

Welch Anova testing Means Equal, allowing Std Devs Not Equal

| F Ratio | DFNum | DFDen | Prob > F |
|---|---|---|---|
| 11.0754 | 5 | 5.8537 | 0.0059* |

Examining the Levene test for equal variances, we have a p-value of 0.0084 rejecting the null hypothesis of all variances being equal. However, the results show a warning that the sample size is relatively small. So, in this situation, we will disregard these results.

# References

Cravens, D. W., R. B. Woodruff, and J. C. Stamper. (January 1972). "An Analytical Approach for Evaluating Sales Territory Performance." *Journal of Marketing*, Vol. 36, No. 1, 31–37.

Eysenck, M.W. (1974). Age differences in incidental learning. *Developmental Psychology*, Vol. 10, No. 6, 936–941.

Marquardt, D.W. (1980). "You Should Standardize the Predictor Variables in Your Regression Models." Discussion of "A Critique of Some Ridge Regression Methods," by G. Smith and F. Campbell. *Journal of the American Statistical Association*, Vol. 75, No. 369, 87–91.

Snee, R. D. (1973). "Some Aspects of Nonorthogonal Data Analysis: Part I. Developing Prediction Equations." *Journal of Quality Technology*, Vol. 5, No. 2, 67–79.

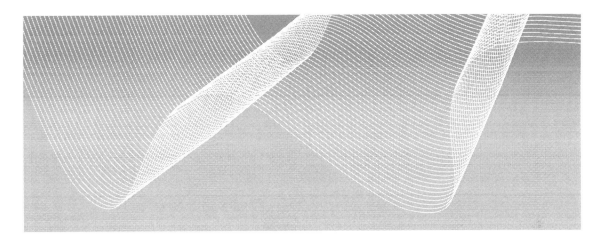

# Chapter 5

## Logistic Regression

**Figure 5.1  A Framework for Multivariate Analysis**

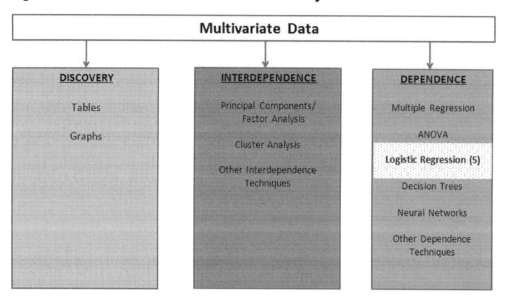

# Dependence Technique: Logistic Regression

Logistic regression, as shown in our multivariate analysis framework in Figure 5.1, is one of the dependence techniques in which the dependent variable is discrete and, more specifically, binary. That is, it takes on only two possible values. Here are some examples: Will a credit card applicant pay off a bill or not? Will a mortgage applicant default? Will someone who receives a direct mail solicitation respond to the solicitation? In each of these cases, the answer is either "yes" or "no." Such a categorical variable cannot directly be used as a dependent variable in a regression. But a simple transformation solves the problem: Let the dependent variable Y take on the value 1 for "yes" and 0 for "no."

Because Y takes on only the values 0 and 1, we know $E[Y_i] = 1*P[Y_i=1] + 0*P[Y_i=0] = P[Y_i=1]$ . But from the theory of regression, we also know that $E[Y_i] = a + b*X_i$. (Here we use simple regression, but the same holds true for multiple regression.) Combining these two results, we have $P[Y_i=1] = a + b*X_i$. We can see that, in the case of a binary dependent variable, the regression may be interpreted as a probability. We then seek to use this regression to estimate the probability that Y takes on the value 1. If the estimated probability is high enough, say above 0.5, then we predict 1; conversely, if the estimated probability of a 1 is low enough, say below 0.5, then we predict 0.

# The Linear Probability Model (LPM)

When linear regression is applied to a binary dependent variable, it is commonly called the Linear Probability Model (LPM). Traditional linear regression is designed for a continuous dependent variable, and is not well-suited to handling a binary dependent variable. Three primary difficulties arise in the LPM. First, the predictions from a linear regression do not necessarily fall between zero and one. What are we to make of a predicted probability greater than one? How do we interpret a negative probability? A model that is capable of producing such nonsensical results does not inspire confidence.

Second, for any given predicted value of y (denoted $\hat{y}$), the residual (resid= y - $\hat{y}$) can take only two values. For example, if $\hat{y} = 0.37$, then the only possible values for the residual are resid= -0.37 or resid = 0.63 (= $1 - 0.37$), because it has to be the case that $\hat{y}$ + resid equals zero or one. Clearly, the residuals will not be normal. Plotting a graph of $\hat{y}$ versus resid will produce not a nice scatter of points, but two parallel lines. The reader should verify this assertion by running such a regression and making the requisite scatterplot. A further implication of the fact that the residual can take on only two values for any $\hat{y}$ is that the residuals are heteroscedastic. This violates the linear regression assumption of homoscedasticity (constant variance). The estimates of the standard errors of the regression coefficients will not be stable and inference will be unreliable.

Third, the linearity assumption is likely to be invalid, especially at the extremes of the independent variable. Suppose we are modeling the probability that a consumer will pay back a $10,000 loan as a function of his/her income. The dependent variable is binary, 1 = the consumer pays back the loan, 0 = the consumer does not pay back the loan. The independent variable is income, measured in dollars. A consumer whose income is $50,000 might have a probability of 0.5 of paying back the loan. If the consumer's income is increased by $5,000, then the probability of paying back the loan might increase to 0.55, so that every $1,000 increase in income increases the probability of paying back the loan by 1%. A person with an income of $150,000 (who can pay the loan back very easily) might have a probability of 0.99 of paying back the loan. What happens to this probability when the consumer's income is increased by $5,000? Probability cannot increase by 5%, because then it would exceed 100%; yet according to the linearity assumption of linear regression, it must do so.

# The Logistic Function

A better way to model $P[Y_i=1]$ would be to use a function that is not linear, one that increases slowly when $P[Y_i=1]$ is close to zero or one, and that increases more rapidly in between. It would have an "S" shape. One such function is the logistic function

$$G(z) = \frac{1}{1+e^{-z}} = \frac{e^z}{1+e^z}$$

whose cumulative distribution function is shown in Figure 5.2.

**Figure 5.2  The Logistic Function**

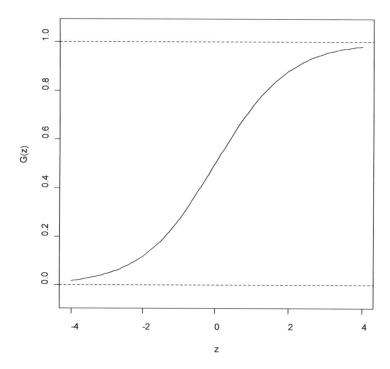

Another useful representation of the logistic function is

$$1-G(z) = \frac{e^{-z}}{1+e^{-z}}$$

Recognize that the y-axis, G(z), is a probability and let G(z) = π, the probability of the event occurring. We can form the odds ratio (the probability of the event occurring divided by the probability of the event not occurring) and do some simplifying:

$$\frac{\pi}{1-\pi} = \frac{G(z)}{1-G(z)} = \frac{\dfrac{1}{1+e^{-z}}}{\dfrac{e^{-z}}{1+e^{-z}}} = \frac{1}{e^{-z}} = e^{z}$$

Consider taking the natural logarithm of both sides. The left side will become log[ $\pi / (1 - \pi)$] and the log of the odds ratio is called the logit. The right side will become $z$ (since log( $e^{z}$ ) = z) so that we have the relation

$$\log\left[\frac{\pi}{1-\pi}\right] = z$$

and this is called the logit transformation.

If we model the logit as a linear function of X (*i.e.*, let z = $\beta_0 + \beta_1 X$ ), then we have

$$\log\left[\frac{\pi}{1-\pi}\right] = \beta_0 + \beta_1 X$$

We could estimate this model by linear regression and obtain estimates $b_0$ of $\beta_0$ and $b_1$ of $\beta_1$ if only we knew the log of the odds ratio for each observation. Since we do not know the log of the odds ratio for each observation, we will use a form of nonlinear regression called logistic regression to estimate the model below:

$$E[Y_i] = \pi_i = G\left(\beta_0 + \beta_1 X_i\right) = \frac{1}{1+e^{-\beta_0 - \beta_1 X_i}}$$

In so doing, we obtain the desired estimates $b_0$ of $\beta_0$ and $b_1$ of $\beta_1$. The estimated probability for an observation $X_i$ will be

$$P\left[Y_i = 1\right] = \hat{\pi}_i = \frac{1}{1+e^{-b_0 - b_1 X_i}}$$

and the corresponding estimated logit will be

$$\log\left[\frac{\hat{\pi}}{1-\hat{\pi}}\right] = b_0 + b_1 X$$

which leads to a natural interpretation of the estimated coefficient in a logistic regression: $b_1$ is the estimated change in the logit (log odds) for a one-unit change in X.

# Example: toylogistic.jmp

To make these ideas concrete, suppose we open a small data set toylogistic.jmp, containing students' midterm exam scores (MidtermScore) and whether the student passed the class (PassClass=1 if pass, PassClass=0 if fail). A passing grade for the midterm is 70. The first thing to do is create a dummy variable to indicate whether the student passed the midterm: PassMidterm = 1 if MidtermScore $\geq$ 70 and PassMidterm = 0 otherwise:

Select **Cols**→**New Column** to open the New Column dialog box. In the Column Name text box, for our new dummy variable, type PassMidterm. Click the drop-down box for modeling type and change it to **Nominal**. Click the drop-down box for Column Properties and select **Formula**. The Formula dialog box appears. Under Functions, click **Conditional**→**If**. Under Table Columns, click **MidtermScore** so that it appears in the top box to the right of the If. Under Functions, click **Comparison Analyze**→**Distributions** **"a>=b"**. In the formula box to the right of >=, enter 70. Press the **Tab** key. Click in the box to the right of the ⇒, and enter the number 1. Similarly, enter 0 for the else clause. The Formula dialog box should look like Figure 5.3.

**Figure 5.3 Formula Dialog Box**

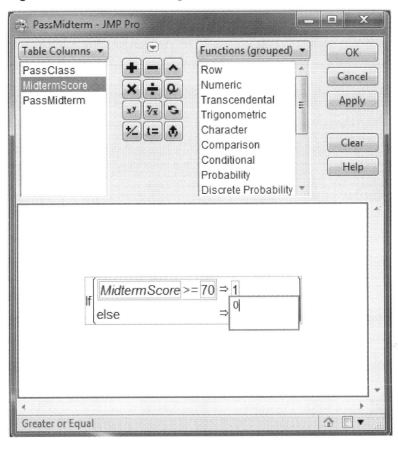

Select **OK**→**OK**.

First, let us use a traditional contingency table analysis to determine the odds ratio. Make sure that both PassClass and PassMidterm are classified as nominal variables. Right-click in the data grid of the column PassClass and select **Column Info**. Click the black triangle next to Modeling Type and select **Nominal**→**OK**. Do the same for PassMidterm.

Select **Tables**→**Tabulate** to open the Control Panel. It shows the general layout for a table. Drag PassClass into the Drop zone for columns and select **Add Grouping Columns**. Now that data have been added, the words Drop zone for rows will no longer be visible, but the Drop zone for rows will still be in the lower left panel of the table. See Figure 5.4.

**Figure 5.4  Control Panel for Tabulate**

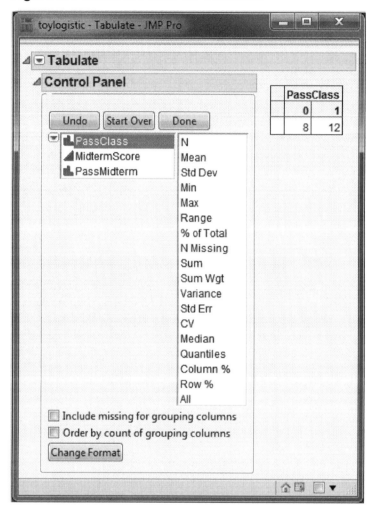

Drag PassMidterm to the panel immediately to the left of the 8 in the table. Select **Add Grouping Columns**. Click **Done**. A contingency table identical to Figure 5.5 will appear.

**Figure 5.5 Contingency Table from toydataset.jmp**

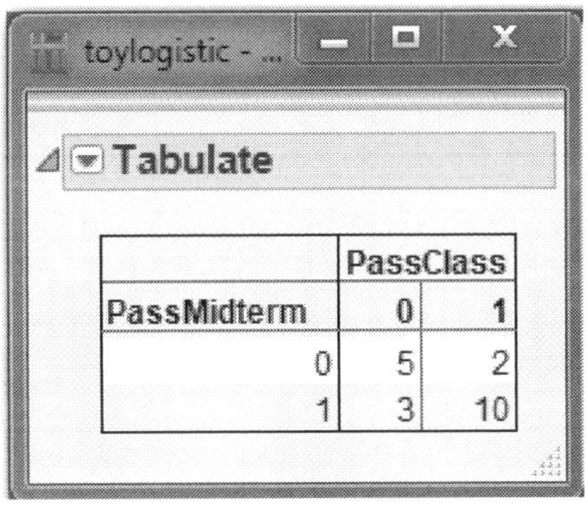

toylogistic - ...

Tabulate

| | PassClass | |
|---|---|---|
| PassMidterm | 0 | 1 |
| 0 | 5 | 2 |
| 1 | 3 | 10 |

The probability of passing the class when you did not pass the midterm is

$$P(PassClass=1)|P(PassMidterm=0) = 2/7$$

The probability of not passing the class when you did not pass the midterm is

$$P(PassClass=0)|P(PassMidterm=0) = 5/7$$

(similar to row percentages). The odds of passing the class given that you have failed the midterm are

$$\frac{P(PassClass=1)\,|\,P(PassMidterm=0)}{P(PassClass=0)\,|\,P((PassMidterm=0)} = \frac{2/7}{5/7} = \frac{2}{5}$$

Similarly, we calculate the odds of passing the class given that you have passed the midterm as:

$$\frac{P(PassClass=1)\,|\,P(PassMidterm=1)}{P(PassClass=0)\,|\,P(PassMidterm=1)} = \frac{10/13}{3/13} = \frac{10}{3}$$

Of the students that did pass the midterm, the odds are the number of students that pass the class divided by the number of students that did not pass the class.

In the above paragraphs, we spoke only of odds. Now let us calculate an odds ratio. It is important to note that this can be done in two equivalent ways. Suppose we want to know the odds ratio of passing the class by comparing those who pass the midterm

(PassMidterm=1 in the numerator) to those who fail the midterm (PassMidterm=0 in the denominator). The usual calculation leads to:

$$\frac{\text{Odds of passing the class; given passed the Midterm}}{\text{Odds of passing the class; given failed the Midterm}} = \frac{10/3}{2/5} = \frac{50}{6} = 8.33.$$

which has the following interpretation: the odds of passing the class are 8.33 times the odds of failing the course if you pass the midterm. This odds ratio can be converted into a probability. We know that $P(Y=1)/P(Y=0)=8.33$; and by definition, $P(Y=1)+P(Y=0)=1$. So solving two equations in two unknowns yields $P(Y=0) = (1/(1+8.33)) = (1/9.33)= 0.1072$ and $P(Y=1) = 0.8928$. As a quick check, observe that $0.8928/0.1072=8.33$. Note that the log-odds are $\ln(8.33) = 2.120$. Of course, the user doesn't have to perform all these calculations by hand; JMP will do them automatically. When a logistic regression has been run, simply clicking the red triangle and selecting **Odds Ratios** will do the trick.

Equivalently, we could compare those who fail the midterm (PassMidterm=0 in the numerator) to those who pass the midterm (PassMidterm=1 in the denominator) and calculate:

$$\frac{\text{Odds of passing the class; given failed the Midterm}}{\text{Odds of passing the class; given passed the Midterm}} = \frac{2/5}{10/3} = \frac{6}{50} = \frac{1}{8.33} = 0.12.$$

which tells us that the odds of failing the class are 0.12 times the odds of passing the class for a student who passes the midterm. Since $P(Y = 0) = 1 - \pi$ (the probability of failing the midterm) is in the numerator of this odds ratio, we must interpret it in terms of the event failing the midterm. It is easier to interpret the odds ratio when it is less than 1 by using the following transformation: $(OR - 1)*100\%$. Compared to a person who passes the midterm, a person who fails the midterm is 12% as likely to pass the class; or equivalently, a person who fails the midterm is 88% less likely, $(OR - 1)*100\% = (0.12 - 1)*100\% = -88\%$, to pass the class than someone who passed the midterm. Note that the log-odds are $\ln(0.12) = -2.12$.

The relationships between probabilities, odds (ratios), and log-odds (ratios) are straightforward. An event with a small probability has small odds, and also has small log-odds. An event with a large probability has large odds and also large log-odds. Probabilities are always between zero and unity; odds are bounded below by zero but can be arbitrarily large; log-odds can be positive or negative and are not bounded, as shown in Figure 5.6. In particular, if the odds ratio is 1 (so the probability of either event is 0.50), then the log-odds equal zero. Suppose $\pi = 0.55$, so the odds ratio $0.55/0.45 = 1.222$. Then we say that the event in the numerator is $(1.222-1) = 22.2\%$ more likely to occur than the event in the denominator.

# Odds Ratios in Logistic Regression

Different software applications adopt different conventions for handling the expression of odds ratios in logistic regression. By default, JMP = uses the "log odds of 0/1" convention, which puts the 0 in the numerator and the 1 in the denominator. This is a consequence of the sort order of the columns, which we will address shortly.

**Figure 5.6  Ranges of Probabilities, Odds, and Log-odds**

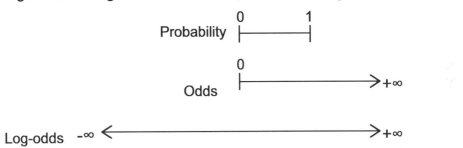

To see the practical importance of this, rather than compute a table and perform the above calculations, we can simply run a logistic regression. It is important to make sure that PassClass is nominal and that PassMidterm is continuous. If PassMidterm is nominal, JMP will fit a different but mathematically equivalent model that will give different (but mathematically equivalent) results. The scope of the reason for this is beyond this book, but, in JMP, interested readers can consult **Help→Books→Modeling and Multivariate Methods** and refer to Appendix A.

If you have been following along with the book, both variables ought to be classified as nominal, so PassMidterm needs to be changed to continuous. Right-click in the column PassMidterm in the data grid and select **Column Info**. Click the black triangle next to Modeling Type and select **Continuous**, and then click **OK**.

Now that the dependent and independent variables are correctly classified as Nominal and Continuous, respectively, let's run the logistic regression:

From the top menu, select **Analyze→Fit Model**. Select **PassClass→Y**. Select **PassMidterm→Add**. The Fit Model dialog box should now look like Figure 5.7. Click **Run**.

### Figure 5.7  Fit Model Dialog Box

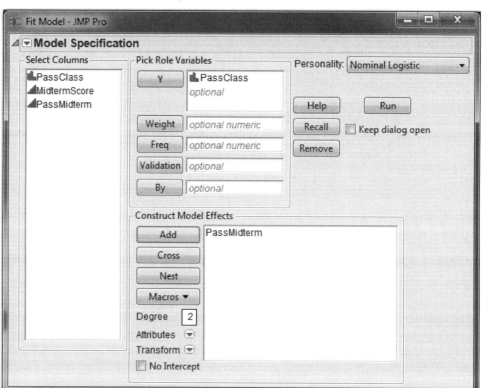

Figure 5.8 displays the logistic regression results.

## Figure 5.8  Logistic Regression Results for toylogistic.jmp

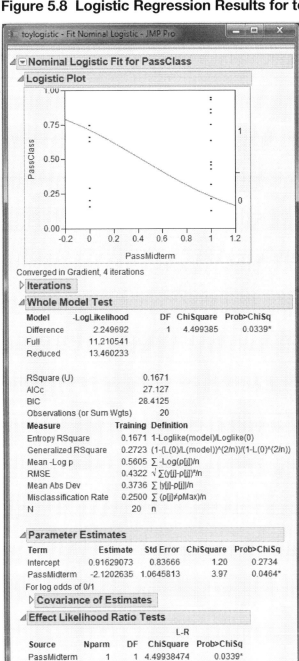

Examine the parameter estimates in Figure 5.8. The intercept is 0.91629073, and the slope is -2.1202635. The slope gives the expected change in the logit for a one-unit change in the independent variable (*i.e.*, the expected change on the log of the odds ratio). However, if we simply exponentiate the slope (*i.e.*, compute ) $e^{-2.1202635} = 0.12$, then we get the 0/1 odds ratio.

There is no need for us to exponentiate the coefficient manually. JMP will do this for us:

Click the red triangle and click **Odds Ratios**. The Odds Ratios tables are added to the JMP output as shown in Figure 5.9.

**Figure 5.9 Odds Ratios Tables Using the Nominal Independent Variable PassMidterm**

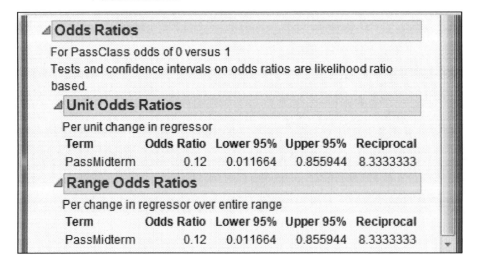

Unit Odds Ratios refers to the expected change in the odds ratio for a one-unit change in the independent variable. Range Odds Ratios refers to the expected change in the odds ratio when the independent variable changes from its minimum to its maximum. Since the present independent variable is a binary 0-1 variable, these two definitions are the same. We get not only the odds ratio, but a confidence interval, too. Notice the right-skewed confidence interval; this is typical of confidence intervals for odds ratios.

To change from the default convention (log odds of 0/1, which puts the 0 in the numerator and the 1 in the denominator, in the data table), right-click to select the name of the PassClass column. Under Column Properties, select **Value Ordering**. Click on the value **1** and click **Move Up** as in Figure 5.10.

**Figure 5.10 Changing the Value Order**

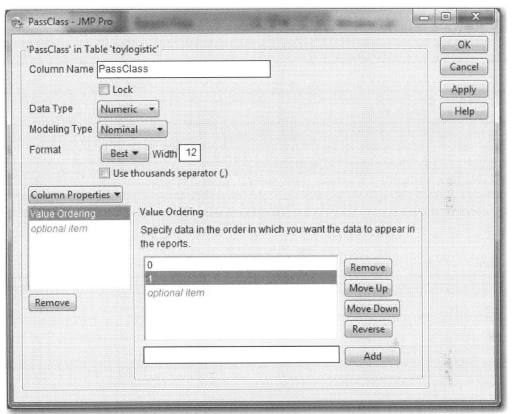

Then, when you re-run the logistic regression, although the parameter estimates will not change, the odds ratios will change to reflect the fact that the 1 is now in the numerator and the 0 is in the denominator.

The independent variable is not limited to being only a nominal (or ordinal) dependent variable; it can be continuous. In particular, let's examine the results using the actual score on the midterm, with MidtermScore as an independent variable:

Select **Analyze→Fit Model**. Select **PassClass→Y** and then select **MidtermScore→Add**. Click **Run**.

This time the intercept is 25.6018754, and the slope is -0.3637609. So we expect the log-odds to decrease by 0.3637609 for every additional point scored on the midterm, as shown in Figure 5.11.

**Figure 5.11  Parameter Estimates**

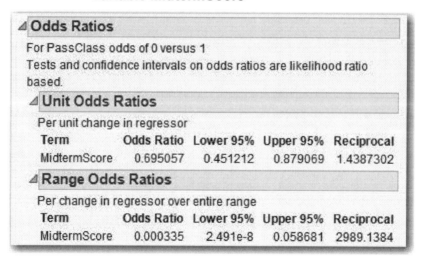

To view the effect on the odds ratio itself, as before click the red triangle and click **Odds Ratios**. Figure 5.12 displays the Odds Ratios tables.

**Figure 5.12  Odds Ratios Tables Using the Continuous Independent Variable MidtermScore**

| ⊿ Odds Ratios | | | |
|---|---|---|---|
| For PassClass odds of 0 versus 1 | | | |
| Tests and confidence intervals on odds ratios are likelihood ratio based. | | | |

| ⊿ Unit Odds Ratios | | | |
|---|---|---|---|
| Per unit change in regressor | | | |
| Term | Odds Ratio | Lower 95% | Upper 95% | Reciprocal |
| MidtermScore | 0.695057 | 0.451212 | 0.879069 | 1.4387302 |

| ⊿ Range Odds Ratios | | | |
|---|---|---|---|
| Per change in regressor over entire range | | | |
| Term | Odds Ratio | Lower 95% | Upper 95% | Reciprocal |
| MidtermScore | 0.000335 | 2.491e-8 | 0.058681 | 2989.1384 |

For a one-unit increase in the midterm score, the new odds ratio will be 69.51% of the old odds ratio. Or, equivalently, we expect to see a 30.5% reduction in the odds ratio (0.695057 – 1)*(100%=-30.5%). For example, suppose a hypothetical student has a midterm score of 75%. The student's log odds of failing the class would be 25.6018754 – 0.3637609*75 = -1.680192. So the student's odds of failing the class would be exp(-1.680192) = 0.1863382. That is, the student is much more likely to pass than fail. Converting odds to probabilities (0.1863328/(1+0.1863328) = 0.157066212786159), we see that the student's probability of failing the class is 0.15707, and the probability of passing the class is 0.84293. Now, if the student's score increased by one point to 76, then the log odds of failing the class would be 25.6018754 – 0.3637609*76 = -2.043953. Thus, the student's odds of failing the class become exp(-2.043953)= 0.1295157. So, the probability of passing the class would rise to 0.885334, and the probability of failing the class would fall to 0.114666. With respect to the Unit Odds Ratio, which equals 0.695057, we see that a one-unit increase in the test score changes the odds ratio from 0.1863382 to 0.1295157. In accordance with the estimated coefficient for the logistic regression, the new odds ratio is 69.5% of the old odds ratio because 0.1295157/0.1863382 = 0.695057.

Finally, we can use the logistic regression to compute probabilities for each observation. As noted, the logistic regression will produce an estimated logit for each observation. These estimated logits can be used, in the obvious way, to compute probabilities for each observation. Consider a student whose midterm score is 70. The student's estimated logit is 25.6018754 – 0.3637609(70) = 0.1386124. Since exp(0.1386129) = 1.148679 = $\pi/(1-\pi)$, we can solve for $\pi$ (the probability of failing) = 0.534597.

We can obtain the estimated logits and probabilities by clicking the red triangle on **Normal Logistic Fit** and selecting **Save Probability Formula**. Four columns will be added to the worksheet: Lin[0], Prob[0], Prob[1], and Most Likely PassClass. For each observation, these give the estimated logit, the probability of failing the class, and the probability of passing the class, respectively. Observe that the sixth student has a midterm score of 70. Look up this student's estimated probability of failing (Prob[0]); it is very close to what we just calculated above. See Figure 5.13. The difference is that the computer carries 16 digits through its calculations, but we carried only six.

**Figure 5.13  Verifying Calculation of Probability of Failing**

| 6/0 | Pass Class | Midterm Score | Lin[0] | Prob[0] | Prob[1] | Most Likely PassClass |
|---|---|---|---|---|---|---|
| 1 | 0 | 62 | 3.048697664 | 0.9547262676 | 0.0452737324 | 0 |
| 2 | 0 | 63 | 2.6849367335 | 0.936131922 | 0.063868078 | 0 |
| 3 | 0 | 64 | 2.3211758029 | 0.9106156911 | 0.0893843089 | 0 |
| 4 | 0 | 65 | 1.9574148724 | 0.8762529099 | 0.1237470901 | 0 |
| 5 | 0 | 66 | 1.5936539419 | 0.8311295662 | 0.1688704338 | 0 |
| 6 | 0 | 70 | 0.1386102197 | 0.5345971803 | 0.4654028197 | 0 |
| 7 | 0 | 72 | -0.588911641 | 0.3568846133 | 0.6431153867 | 1 |
| 8 | 0 | 74 | -1.316433502 | 0.2114122777 | 0.7885877223 | 1 |
| 9 | 1 | 68 | 0.8661320808 | 0.7039402276 | 0.2960597724 | 0 |
| 10 | 1 | 69 | 0.5023711503 | 0.6230163983 | 0.3769836017 | 0 |
| 11 | 1 | 71 | -0.225150711 | 0.4439489049 | 0.5560510951 | 1 |
| 12 | 1 | 73 | -0.952672572 | 0.2783476639 | 0.7216523361 | 1 |

The fourth column (Most Likely PassClass) classifies the observation as either 1 or 0, depending upon whether the probability is greater than or less than 50%. We can observe how well our model classifies all the observations (using this cut-off point of 50%) by producing a confusion matrix: Click the red triangle and click Confusion matrix. Figure 5.14 displays the confusion matrix for our example. The rows of the confusion matrix are the actual classification (that is, whether PassClass is 0 or 1). The columns are the predicted classification from the model (that is, the predicted 0/1 values from that last fourth column using our logistic model and a cutpoint of .50). Correct classifications are along the main diagonal from upper left to lower right. We see that the model has classified 6 students as not passing the class, and actually they did not pass the class. The model also classifies 10 students as passing the class when they actually did. The values on the other diagonal, both equal to 2, are misclassifications. The results of the confusion matrix will be examined in more detail when we discuss model comparison in Chapter 9.

**Figure 5.14 Confusion Matrix**

| ◢ Confusion Matrix | | |
|---|---|---|
| Actual | Predicted | |
| Training | 0 | 1 |
| 0 | 6 | 2 |
| 1 | 2 | 10 |

Of course, before we can use the model, we have to check the model's assumptions, etc. The first step is to verify the linearity of the logit. This can be done by plotting the estimated logit against PassClass. Select **Graph**→**Scatterplot Matrix.** Select **Lin[0]**→**Y, columns**. Select **MidtermScore**→**X**. Click **OK**. As shown in Figure 5.15, the linearity assumption appears to be perfectly satisfied.

**Figure 5.15 Scatterplot of Lin[0] and MidtermScore**

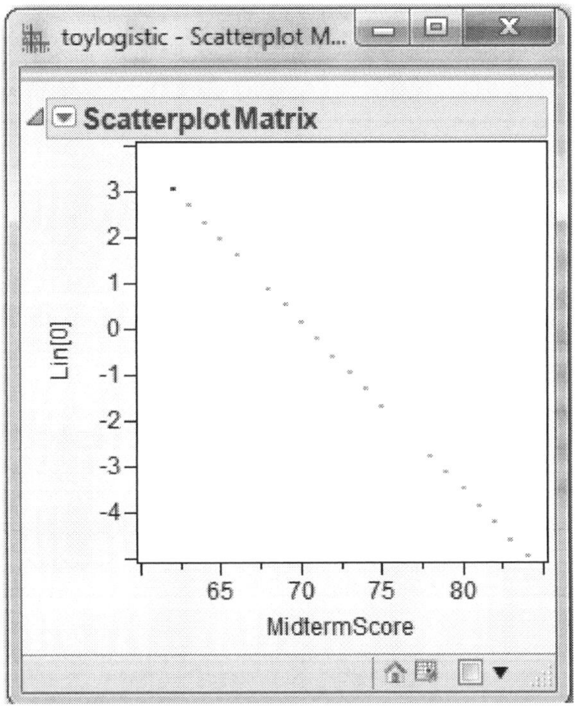

The analog to the ANOVA F-test for linear regression is found under the Whole Model Test, shown in Figure 5.16, in which the Full and Reduced models are compared. The null hypothesis for this test is that all the slope parameters are equal to zero. Since Prob>ChiSq is 0.0004, this null hypothesis is soundly rejected. For a discussion of other statistics found here, such as BIC and Entropy RSquare, see the **JMP Help.**

**Figure 5.16  Whole Model Test for the Toylogistic Data Set**

| Model | -LogLikelihood | DF | ChiSquare | Prob>ChiSq |
|---|---|---|---|---|
| Difference | 6.264486 | 1 | 12.52897 | 0.0004* |
| Full | 7.195748 | | | |
| Reduced | 13.460233 | | | |

| | |
|---|---|
| RSquare (U) | 0.4654 |
| AICc | 19.0974 |
| BIC | 20.383 |
| Observations (or Sum Wgts) | 20 |

*Whole Model Test*

The next important part of model checking is the Lack of Fit test. See Figure 5.17. It compares the model actually fitted to the saturated model. The saturated model is a model generated by JMP that contains as many parameters as there are observations. So it fits the data very well. The null hypothesis for this test is that there is no difference between the estimated model and the saturated model. If this hypothesis is rejected, then more variables (such as cross-product or squared terms) need to be added to the model. In the present case, as can be seen, Prob>ChiSq=0.7032. We can therefore conclude that we do not need to add more terms to the model.

**Figure 5.17  Lack of Fit Test for Current Model**

| Source | DF | -LogLikelihood | ChiSquare |
|---|---|---|---|
| Lack Of Fit | 18 | 7.1957477 | 14.3915 |
| Saturated | 19 | 0.0000000 | Prob>ChiSq |
| Fitted | 1 | 7.1957477 | 0.7032 |

*Lack Of Fit*

# A Logistic Regression Statistical Study

Let's turn now to a more realistic data set with several independent variables. During this discussion, we will also present briefly some of the issues that should be addressed and some of the thought processes during a statistical study.

Cellphone companies are very interested in determining which customers might switch to another company; this is called "churning." Predicting which customers might be about to churn enables the company to make special offers to these customers, possibly stemming their defection. Churn.jmp contains data on 3333 cellphone customers, including the variable Churn (0 means the customer stayed with the company and 1 means the customer left the company).

Before we can begin constructing a model for customer churn, we need to discuss model building for logistic regression. Statistics and econometrics texts devote entire chapters to this concept. In several pages, we can only sketch the broad outline. The first thing to do is make sure that the data are loaded correctly. Observe that Churn is classified as Continuous; be sure to change it to Nominal. One way is to right-click in the Churn column in the data table, select **Column Info**, and under Modeling Type, click **Nominal**. Another way is to look at the list of variables on the left side of the data table, find Churn, click the blue triangle (which denotes a continuous variable), and change it to **Nominal** (the blue triangle then becomes a red histogram). Make sure that all binary variables are classified as Nominal. This includes Intl_Plan, VMail_Plan, E_VMAIL_PLAN, and D_VMAIL_PLAN. Should Area_Code be classified as Continuous or Nominal? (Nominal is the correct answer!) CustServ_Call, the number of calls to customer service, could be treated as either continuous or nominal/ordinal; we treat it as continuous.

When building a linear regression model and the number of variables is not so large that this cannot be done manually, one place to begin is by examining histograms and scatterplots of the continuous variables, and crosstabs of the categorical variables as discussed in Chapter 3. Another very useful device as discussed in Chapter 3 is the scatterplot/correlation matrix, which can, at a glance, suggest potentially *useful* independent variables that are correlated with the dependent variable. The scatterplot/correlation matrix approach cannot be used with logistic regression, which is nonlinear, but a method similar in spirit can be applied.

We are now faced with a similar situation that was discussed in Chapter 4. Our goal is to build a model that follows the principle of parsimony—that is, a model that explains as much as possible of the variation in Y and uses as few significant independent variables as possible. However, now with multiple logistic regression, we are in a nonlinear situation. We have four approaches that we could take. We briefly list and discuss each of these approaches and some of their advantages and disadvantages:

■ **Include all the variables.** In this approach you just input all the independent variables into the model. An obvious advantage of this approach is that it is fast and easy. However, depending on the data set, most likely several independent variables will be insignificantly related to the dependent variable. Including variables that are not significant can cause severe problems—weakening the interpretation of the coefficients and lessening the prediction accuracy of the model. This approach definitely does not follow the principle of parsimony, and it can cause numerical problems for the nonlinear solver that may lead to a failure to obtain an answer.

- **Bivariate method.** In this approach, you search for independent variables that may have predictive value for the dependent variable by running a series of bivariate logistic regressions; *i.e.*, we run a logistic regression for each of the independent variables, searching for "significant" relationships. A major advantage of this approach is that it is the one most agreed upon by statisticians (Hosmer and Lemeshow, 2001). On the other hand, this approach is not automated, is very tedious, and is limited by the analyst's ability to run the regressions. That is, it is not practical with very large data sets. Further, it misses interaction terms, which, as we shall see, can be very important.

- **Stepwise.** In this approach, you would use the Fit Model platform, change the Personality to Stepwise and Direction to Mixed. The Mixed option is like Forward Stepwise, but variables can be dropped after they have been added. An advantage of this approach is that it is automated; so, it is fast and easy. The disadvantage of stepwise is that it could lead to possible interpretation and prediction errors depending on the data set. However, using the Mixed option, as opposed to the Forward or Backward Direction option, tends to lessen the magnitude and likelihood of these problems.

- **Decision Trees.** A Decision Tree is a data mining technique that can be used for variable selection and will be discussed in Chapter 8. The advantage of using the decision tree technique is that it is automated, fast, and easy to run. Further, it is a popular variable reduction approach taken by many data mining analysts (Pollack, 2008). However, somewhat like the stepwise approach, the decision tree approach could lead to some statistical issues. In this case, significant variables identified by a decision tree are very sample-dependent. These issues will be discussed further in Chapter 8.

No one approach is a clear cut winner. Nevertheless, we do not recommend using the "Include all the variables" approach. If the data set is too large and/or you do not have the time, we recommend that you run both the stepwise and decision trees models and compare the results. The data set churn.jmp is not too large, so we will apply the bivariate approach.

It is traditional to choose $\alpha = 0.05$. But in this preliminary stage, we adopt a more lax standard, $\alpha = 0.25$. The reason for this is that we want to include, if possible, a group of variables that individually are not significant but together are significant. Having identified an appropriate set of candidate variables, run a logistic regression including all of them. Compare the coefficient estimates from the multiple logistic regression with the estimates from the bivariate logistic regressions. Look for coefficients that have changed in sign or have dramatically changed in magnitude, as well as changes in significance. Such changes indicate the inadequacy of the simple bivariate models, and confirm the necessity of adding more variables to the model.

Three important ways to improve a model are as follows:

- If the logit appears to be nonlinear when plotted against some continuous variable, one resolution is to convert the continuous variable to a few dummies, say three, that cut the variable at its 25th, 50th, and 75th percentiles.

- If a histogram shows that a continuous variable has an excess of observations at zero (which can lead to nonlinearity in the logit), add a dummy variable that equals one if the continuous variable is zero and equals zero otherwise.

- Finally, a seemingly numeric variable that is actually discrete can be broken up into a handful of dummy variables (*e.g.*, ZIP codes).

Before we can begin modeling, we must first explore the data. With our churn data set, creating and examining the histograms of the continuous variables reveals nothing much of interest, except VMail_Message, which has an excess of zeros. (See the second point in the previous paragraph.) Figure 5.18 shows plots for Intl_Calls and VMail_Message. To produce such plots, select **Analyze→Distribution**, click **Intl_Calls**, and then **Y, Columns** and **OK**. To add the Normal Quantile Plot, click the red arrow next to Intl_Calls and select **Normal Quantile Plot**. Here it is obvious that Intl_Calls is skewed right. We note that a logarithmic transformation of this variable might be in order, but we will not pursue the idea.

**Figure 5.18  Distribution of Intl_Calls and VMail_Message**

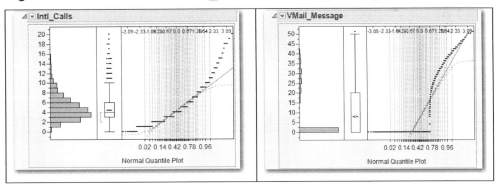

A correlation matrix of the continuous variables (select **Graph→Scatterplot Matrix** and put the desired variables in **Y, Columns**) turns up a curious pattern. Day_Charge and Day_Mins, Eve_Charge and Eve_Mins, Night_Charge and Night_Mins, and Intl_Charge and Intl_Mins all are perfectly correlated. The charge is obviously a linear function of the number of minutes. Therefore, we can drop the Charge variables from our analysis. (We could also drop the "Mins" variables instead; it doesn't matter which one we drop.) If our data set had a very large number of variables, the scatterplot matrix would be too big to comprehend. In such a situation, we would choose groups of variables for which to make scatterplot matrices, and examine those.

A scatterplot matrix for the four binary variables turns up an interesting association. E_VMAIL_PLAN and D_VMAIL_PLAN are perfectly correlated; both have common 1s and where the former has -1, the latter has zero. It would be a mistake to include both of these variables in the same regression (try it and see what happens). Let's delete E_VMAIL_PLAN from the data set and also delete VMail_Plan because it agrees perfectly with E_VMAIL_PLAN: When the former has a "no," the latter has a "-1," and similarly for "yes" and "+1."

Phone is more or less unique to each observation. (We ignore the possibility that two phone numbers are the same but have different area codes.) Therefore, it should not be included in the analysis. So, we will drop Phone from the analysis.

A scatterplot matrix between the remaining continuous and binary variables turns up a curious pattern. D_VMAIL_PLAN and VMailMessage have a correlation of 0.96. They have zeros in common, and where the former has 1s, the latter has numbers. (See again point two in the above paragraph. We won't have to create a dummy variable to solve the problem because D_VMAIL_PLAN will do the job nicely.)

To summarize, we have dropped 7 of the original 23 variables from the data set (Phone, Day_Charge, Eve_Charge, Night_Charge, Intl_Charge, E_VMAIL_PLAN, and VMail_Plan). So there are now 16 variables left, one of which is the dependent variable, Churn. We have 15 possible independent variables to consider.

Next comes the time-consuming task of running several bivariate (two variables, one dependent and one independent) analyses, some of which will be logistic regressions (when the independent variable is continuous) and some of which will be contingency tables (when the independent variable is categorical). In total, we have 15 bivariate analyses to run. What about Area Code? JMP reads it as a continuous variable, but it's really nominal, so make sure to change it from continuous to nominal. Similarly, make sure that D_VMAIL_PLAN is set as a nominal variable, not continuous.

Do *not* try to keep track of the results in your head, or by referring to the 15 bivariate analyses that would fill your computer screen. Make a list of all 15 variables that need to be tested, and write down the test result (*e.g.*, the relevant p-value) and your conclusion (*e.g.*, "include" or "exclude"). This not only prevents simple errors; it is a useful record of your work should you have to come back to it later. There are few things more pointless than conducting an analysis that concludes with a 13-variable logistic regression, only to have some reason to rerun the analysis and now wind up with a 12-variable logistic regression. Unless you have documented your work, you will have no idea why the discrepancy exists or which is the correct regression.

Below we briefly show how to conduct both types of bivariate analyses, one for a nominal independent variable and one for a continuous independent variable. We leave the other 14 to the reader.

Make a contingency table of Churn versus State: Select **Analyze→Fit Y by X**, click Churn (which is nominal) and then click **Y, Response,** click **State** and then click **X, Factor**; and click **OK**. At the bottom of the table of results are the Likelihood Ratio and Pearson tests, both of which test the null hypothesis that **State** does not affect **Churn**, and both of which reject the null. The conclusion is that the variable **State** matters. On the other hand, perform a logistic regression of Churn on VMail_Message: select **Analyze→Fit Y by X**, click **Churn**, click **Y, Response**, and click **VMail_Message** and click **X**, Factor; and click **OK**. Under "Whole Model Test" that Prob>ChiSq, the p-value is less than 0.0001, so we conclude that **VMail_message** affects **Churn**. Remember that for all these tests, we are setting $\alpha$ (probability of Type I error) = 0.25.

In the end, we have 10 candidate variables for possible inclusion in our multiple logistic regression model:

| | | |
|---|---|---|
| State | Intl_Plan | D_VMAIL_PLAN |
| VMail_Message | Day_Mins | Eve_Mins |
| Night_Mins | Intl_Mins | Intl_Calls |
| CustServ_Call | | |

Remember that the first three of these variables (the first row) should be set to nominal, and the rest to continuous. (Of course, leave the dependent variable Churn as nominal!)

Let's run our initial multiple logistic regression with Churn as the dependent variable and the above 10 variables as independent variables:

Select **Analyze→Fit Model→Churn→Y**. Select the above 10 variables (to select variables that are not consecutive, click on each variable while holding down the **Ctrl** key), and click **Add**. Check the box next to Keep dialog open. Click **Run**.

The Whole Model Test lets us know that our included variables have an effect on the Churn and a p-value less than .0001, as shown in Figure 5.19.

**Figure 5.19 Whole Model Test and Lack of Fit for the Churn Data Set**

### Nominal Logistic Fit for Churn

Converged in Gradient, 6 iterations

▷ **Iterations**

#### Whole Model Test

| Model | -LogLikelihood | DF | ChiSquare | Prob>ChiSq |
|---|---|---|---|---|
| Difference | 341.6995 | 59 | 683.3991 | <.0001* |
| Full | 1037.4471 | | | |
| Reduced | 1379.1467 | | | |

| | |
|---|---|
| RSquare (U) | 0.2478 |
| AICc | 2197.13 |
| BIC | 2561.59 |
| Observations (or Sum Wgts) | 3333 |

| Measure | Training | Definition |
|---|---|---|
| Entropy RSquare | 0.2478 | 1-Loglike(model)/Loglike(0) |
| Generalized RSquare | 0.3293 | $(1-(L(0)/L(model))^{(2/n)})/(1-L(0)^{(2/n)})$ |
| Mean -Log p | 0.3113 | $\sum -Log(p[j])/n$ |
| RMSE | 0.3070 | $\sqrt{\sum (y[j]-p[j])^2/n}$ |
| Mean Abs Dev | 0.1883 | $\sum |y[j]-p[j]|/n$ |
| Misclassification Rate | 0.1308 | $\sum (p[j]\ne pMax)/n$ |
| N | 3333 | n |

#### Lack Of Fit

| Source | DF | -LogLikelihood | ChiSquare |
|---|---|---|---|
| Lack Of Fit | 3273 | 1037.4471 | 2074.894 |
| Saturated | 3332 | 0.0000 | Prob>ChiSq |
| Fitted | 59 | 1037.4471 | 1.0000 |

The lack-of-fit test tells us that we have done a good job explaining Churn. From the Lack of Fit, we see that –LogLikelihood for the Full model is 1037.4471. Now, linear regression minimizes the sum of squared residuals. So when you compare two linear regressions, the preferred one has the smaller sum of squared residuals. In the same way, the nonlinear optimization of the logistic regression minimizes the –LogLikelihood (which is equivalent to maximizing the LogLikelihood). So the model with the smaller –LogLikelihood is preferred to a model with a larger –LogLikelihood.

Examining the p-values of the independent variables in the Parameter Estimates, we find that a variable for which Prob>ChiSq is less than 0.05 is said to be significant. Otherwise, it is said to be insignificant, similar to what is practiced in linear regression. The regression output gives two sets of tests, one for the "Parameter Estimates" and another for "Effect Likelihood Ratio Tests." We shall focus on the latter. To see why, consider the State variable, which is really not one variable but many dummy variables. We are not so much interested in whether any particular state is significant or not (which is what the Parameter Estimates tell us) but whether, overall, the collection of state dummy variables is significant. This is what the Effect Likelihood Ratio Tests tells us; the effect of all the state dummies is significant with a "Prob>ChiSq" of 0.0010. True, many of the State dummies are insignificant, but overall State is significant; we will keep this variable as it is. It may prove worthwhile to reduce the number of state dummies into a handful of significant states and small clusters of "other" states that are not significant, but we will not pursue this line of inquiry here.

We can see that all the variables in the model are significant. We may be able to derive some new variables that help improve the model. We will provide two examples of deriving new variables—(1) Converting a continuous variable into discrete variables; (2) Producing interaction variables.

Let us try to break up a continuous variable into a handful of discrete variables. An obvious candidate is CustServ_Call. Look at its distribution in Figure 5.20. Select **Analyze→Distribution**, select **CustServ_Call→Y, Columns**, and click **OK**. Click the red arrow next to CustServ_Call and uncheck **Outlier Box Plot**. Then choose **Histogram Options→Show Counts**.

### Figure 5.20  Histogram of CustServ_Call

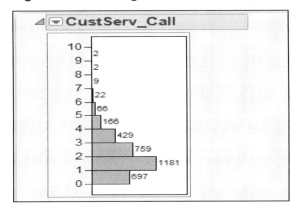

Let's create a new nominal variable called CustServ, so that all the counts for 5 and greater are collapsed into a single cell:

> Select **Cols→New Columns**. For column name type CustServ, for Modeling Type change it to **Nominal** and then click the drop-down arrow for Column Properties and click **Formula**. In the Formula dialog box, select **Conditional→If**. Then, in the top expr, click **CustServ_Call** and type <=4. In

the top then clause, click **CustServ_Call**. For the else clause, type **5**. See Figure 5.21. Click **OK** and click **OK**.

**Figure 5.21  Creating the CustServ Variable**

Now drop the CustServ_Call variable from the Logistic Regression and add the new CustServ nominal variable, which is equivalent to adding some dummy variables. Our new value of -LogLikelihood is 970.6171, which constitutes a very substantial improvement in the model.

Another possible important way to improve a model is to introduce interactions terms, that is, the product of two or more variables. Best practice would be to consult with subject-matter experts and seek their advice. Some thought is necessary to determine meaningful interactions, but it can pay off in substantially improved models. Thinking about what might make a cell phone customer want to switch to another carrier, we have all heard a friend complain about being charged an outrageous amount for making an international call. Based on this observation, we could conjecture that customers who make international calls and who are not on the international calling plan might be more irritated and more likely to churn. A quick bivariate analysis shows that there are more than a few such persons in the data set. Select **Tables**→**Tabulate**, and drag Intl_Plan to Drop zone for columns. Drag Intl_Calls to Drop zone for rows. Click **Add Grouping Columns**. Observe that almost all customers make international calls, but most of them are not on the international plan (which gives cheaper rates for international calls). For example, for the customers who made no international call, all 18 of them were not on the international calling plan. For the customers who made 8 international calls, 106 were not on the international calling plan, and only 10 of them were. There is quite the potential for irritated customers here! This is confirmed by examining the output from the previous logistic regression. The parameter estimate for "Intl_Plan[no]" is positive and significant. This means that when a customer does not have an international plan, the probability is that the churn increases.

Customers who make international calls and don't get the cheap rates are perhaps more likely to churn than customers who make international calls and get cheap rates. Hence, the interaction term Int_Plan*Intl_Mins might be important. To create this interaction term, we have to create a new dummy variable for Intl_Plan, because the present variable is not numeric and cannot be multiplied by Intl_Mins:

> First, click on the Intl_Plan column in the data table to select it. Then select **Cols**→**Recode**. Under **New Value**, where it has **No**, type **0** and right below that where it has **Yes**, type **1**. From the **In Place** drop-down menu, select **New Column** and click OK. The new variable Intl_Plan2 is created. However, it is still nominal. Right-click on this column and under **Column Info**, change the Data Type to **Numeric** and the Modeling Type to **Continuous**. Click **OK**. (This variable has to be continuous so that we can use it in the interaction term, which is created by multiplication; nominal variables cannot be multiplied.)

To create the interaction term:

> Select **Cols**→**New Column** and call the new variable IntlPlanMins. Under Column Properties, click **Formula**. Click **Intl_Plan2**, click on the times sign (**x**) in the middle of the dialog box, click **Intl_Mins** and click **OK**. Click **OK** again.

Now add the variable IntlPlanMins as the 11th independent variable in multiple logistic regression that includes CustServ and run it. The variable IntlPlanMins is significant, and the –LogLikelihood has dropped to 947.1450, as shown in Figure 5.22. This is a substantial drop for adding one variable. Doubtless other useful interaction terms could be added to this model, but we will not further pursue this line of inquiry.

**Figure 5.22 Logistic Regression Results with Interaction Term Added**

### ▲ ⊡ Nominal Logistic Fit for Churn

Converged in Gradient, 6 iterations

▷ **Iterations**

### ▲ Whole Model Test

| Model | -LogLikelihood | DF | ChiSquare | Prob>ChiSq |
|---|---|---|---|---|
| Difference | 432.0016 | 65 | 864.0033 | <.0001* |
| Full | 947.1450 | | | |
| Reduced | 1379.1467 | | | |

Now that we have built an acceptable model, it is time to validate the model. We have already checked the Lack of Fit, but now we have to check linearity of the logit. From the red arrow, click **Save Probability Formula**, which adds four variables to the data set: Lin[0] (which is the logit), Prob[0], Prob[1], and the predicted value of Churn, Most Likely Churn. Now we have to plot the logit against each of the continuous independent variables. The categorical independent variables do not offer much opportunity to reveal nonlinearity (plot some and see this for yourself). All the relationships of the continuous variables can be quickly viewed by generating a scatterplot matrix and then clicking the red triangle and **Fit Line**. Nearly all the red fitted lines are horizontal or near horizontal. For all of the logit vs. independent variable plots, there is no evidence of nonlinearity.

We can also see how well our model is predicting by examining the confusion matrix, which is shown in Figure 5.23.

**Figure 5.23 Confusion Matrix**

### ▲ Confusion Matrix

| Actual | | Predicted |
|---|---|---|
| Training | 0 | 1 |
| 0 | 2749 | 101 |
| 1 | 326 | 157 |

The actual number of churners in the data set is 326+157 = 483. The model predicted a total of 258 (=101+157) churners. The number of bad predictions made by the model is 326+101 = 427, which indicates that 326 that were predicted not to churn actually did churn, and 101 that were predicted to churn did not churn. Further, observe in the Prob[1] column of the data table that we have the probability that any customer will churn. Right-click on this column and select **Sort**. This will sort all the variables in the data set according to the probability of churning. Scroll to the top of the data set. Look at the

Churn column. It has mostly ones and some zeros here at the top, where the probabilities are all above 0.85. Scroll all the way to the bottom and see that the probabilities now are all below 0.01, and the values of Churn are all zero. We really have modeled the probability of churning.

Now that we have built a model for predicting churn, how might we use it? We could take the next month's data (when we do not yet know who has churned) and predict who is likely to churn. Then these customers can be offered special deals to keep them with the company, so that they do not churn.

# References

Hosmer, D. W., and S. Lemeshow. (2001). *Applied Logistic Regression*. 2nd ed. New York: John Wiley & Sons.

Pollack, R. (2008). "Data Mining: Common Definitions, Applications, and Misunderstandings." *Data Mining Methods and Applications (Discrete Mathematics & Its Applications)*. Lawrence, K. D., S. Kudyba, and R. K. Klimberg (Eds.). Boca Raton, FL: Auerbach Publications.

# Exercises

1.  Consider the logistic regression for the toy data set, where $\pi$ is the probability of passing the class:

$$\log\left[\frac{\hat{\pi}}{1-\hat{\pi}}\right] = 25.60188 - 0.363761\, MidtermScore$$

Consider two students, one who scores 67% on the midterm and one who scores 73% on the midterm. What are the odds that each fails the class? What is the probability that each fails the class?

2.  Consider the first logistic regression for the Churn data set, the one with 10 independent variables. Consider two customers, one with an international plan and one without. What are the odds that each churns? What is the probability that each churns?

3. We have already found that the interaction term IntlPlanMins significantly improves the model. Find another interaction term that does so.

4. Without deriving new variables such as CustServ or creating interaction terms such as IntlPlanMins, use a stepwise method to select variables for the Churn data set. Compare your results to the bivariate method used in the chapter; pay particular attention to the fit of the model and the confusion matrix.

5. Use the Freshmen1.jmp data set and build a logistic regression model to predict whether a student returns. Perhaps the continuous variables Miles from Home and Part Time Work Hours do not seem to have an effect. See whether turning them into discrete variables makes a difference. (*E.g.*, turn Miles from Home into some dummy variables, 0–20 miles, 21–100 miles, more than 100 miles.)

# Chapter 6

# Principal Components Analysis

**Figure 6.1 A Framework for Multivariate Analysis**

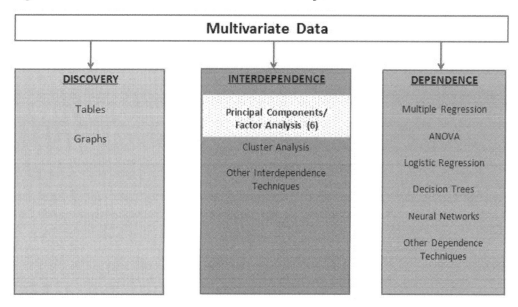

Principal Component Analysis (PCA) is an exploratory multivariate technique with two overall objectives. One objective is "dimension reduction"— *i.e.*, to turn a collection of, say, 100 variables into a collection of 10 variables that retain almost all the information that was contained in the original 100 variables. The other objective is to discover the structure in the relationships between the variables. As shown in Figure 6.1, PCA is a technique that does not require a dependent variable.

PCA analyzes the structure of the interrelationships (correlations) among the variables by defining a set of common underlying dimensions called components or factors. PCA achieves this goal by eliminating unnecessary correlations between variables. To the extent that two variables are correlated, there are really three sources of variation: variation unique to the first variable, variation unique to the second variable, and variations common to the two variables. The method of PCA transforms the two variables into uncorrelated variables (with no common variation) that still preserve two sources of unique variation so that the total variation in the two variables remains the same.

To see how this works, let's examine a small data table toyprincomp.jmp, which contains the three variables, x, y, and z. Before performing the PCA, let's first examine the correlations and scatterplot matrix. To produce them:

> Select **Analyze**→**Multivariate Methods**→**Multivariate**. For the Select Columns option in the Multivariate and Correlations dialog box, click **X**, hold down the shift key, and click **Y** to select all three variables. Click **Y,Columns**,

and all the variables will appear in the white box to the right of **Y, Columns**. Click **OK**.

As shown in Figure 6.2, the correlation matrix and corresponding scatterplot matrix will be generated. There appear to be several strong correlations among all the variables.

**Figure 6.2  Correlations and Scatterplot Matrix for the toyprincomp.xls Data Set**

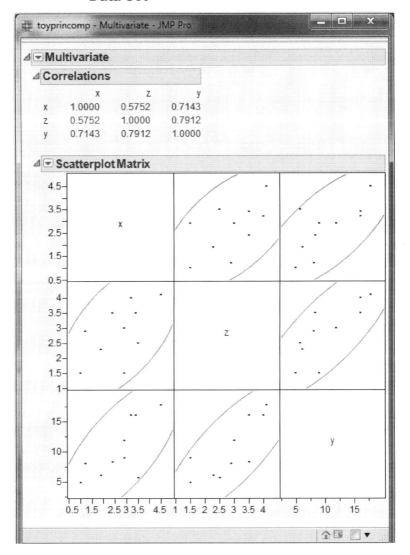

We will treat x and z as a pair of variables on which to apply PCA; we'll reserve y for use as a dependent variable for running regressions. Since we have two variables, x and z, we will create two principal components.

Open toyprincomp.xls in JMP. Click **Analyze→Multivariate Methods→Principal Components**. In the Principal Components dialog box, click **x**, hold down the shift key, and click **z** to select both variables. Then click **Y, Columns** as shown in Figure 6.3. Click **OK**.

### Figure 6.3  Principal Components Dialog Box

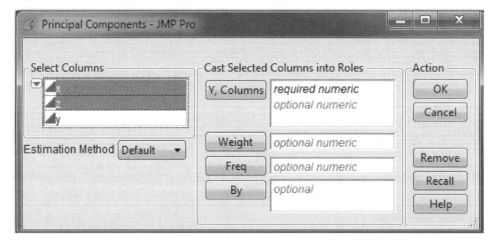

Three principal component summary plots, as in Figure 6.4, will appear. These plots show how the principal components explain the variation in the data. The left-most graph is a listing and bar chart of the eigenvalue and percentage of variation accounted for by each principal component. (In this case, with only two principal components, the graph is not too informative.)

**Figure 6.4 Principal Components Summary Plots**

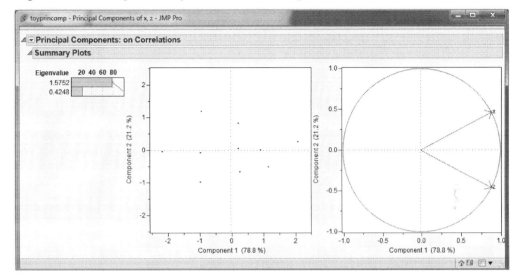

The first eigenvalue, 1.5752, is much larger than the second eigenvalue, 0.4248. This suggests that the first principal component, Prin1, is much more important (in terms of explaining the variation in the pair of variables) than the second principal component, Prin2. Additionally, the bar for the percentage of variation accounted for by Prin1 is about 80%. (As we can see in the other two plots, Prin1 accounted for 78.8% and Prin2 accounted for 21.2%.)

The second graph is a scatter plot of the two principal components; this is called a Score Plot. Notice that the scatter is approximately flat, indicating a lack of correlation between the principal components. A correlation between the principal components would be indicated by a scatter with a positive or negative slope.

The third graph is called a Loadings Plot, and it shows the contribution made to each principal component (the principal components are the axes of the graph) by the original variables (which are the directed points on the graph). In this case, reading from the plot, we see that x and z each contribute about 0.90 to Prin1, while x contributes about a positive 0.50 to Prin2 and z contributes about a negative 0.50 to Prin2. This graph is not particularly useful when there are only two variables. When there are many variables, it is possible to see which variables group together in the principal component space; examples of this will be presented in a later section.

Linear combinations of x and z have been formed into Prin1 and Prin2. To show that Prin1 and Prin2 represent the same overall correlation as x and z, we use the dependent variable y and run two regressions: First regress y on x and z. Then regress y on Prin1 and Prin2. (The variables Prin1 and Prin2 can be created by clicking the red triangle, clicking **Save Principal Components**, and then entering **2** as the number of components to save. Click **OK**.) Both regressions have the exact same $R^2$ and Root Mean Square Error (or $s_e$ as shown in Table 6.1).

To see the true value of the principal components method, now run a regression of y on Prin1, excluding Prin2. We get almost the same $R^2$ (which is slightly lower as shown in Table 6.1), which implies that the predicted y values from both models are essentially the same. We also see a drop in the Root Mean Square Error, which implies a tighter prediction interval. If our interest is not in interpreting coefficients but simply in predicting by use of principal components, we get practically the same prediction with a tighter prediction interval. The attraction of PCA is obvious.

**Table 6.1 Regression Results**

|  | $R^2$ | Root Mean Square Error ($s_e$) |
|---|---|---|
| y\|x, z | 0.726336 | 2.811032 |
| y\| Prin1, Prin2 | 0.726336 | 2.811032 |
| y\| Prin1 | 0.719376 | 2.662706 |

# Principal Component

The method of principal components or PCA works by transforming a set of k correlated variables into a set of k uncorrelated variables that are called, not coincidentally, principal components or simply components. The first principal component (Prin1) is a linear combination of the original variables:

$$Prin1 = w_{11}x_1 + w_{12}x_2 + w_{13}x_3 + \ldots + w_{1k}x_k$$

Prin1 accounts for the most variation in the data set. The second principal component is uncorrelated with the first principal component, and accounts for the second-most variation in the data set. Similarly, Prin3 is uncorrelated with Prin1 and Prin2 and accounts for the third-most variation in the data set. The general form for the *i*th principal component is

$$Prin(i) = w_{i1}x_1 + w_{i2}x_2 + w_{i3}x_3 + \ldots + w_{ik}x_k$$

where $w_{i1}$ is the weight of the first variable in the *i*th principal component.

(Note: A related multivariate technique is factor analysis. Although superficially quite similar, the techniques are quite different, and have different purposes. First, the PCA approach accounts for all the variance, common and unique; factor analysis analyzes only common variance. Further, with PCA, each principal component is viewed as a weighted combination of input variables. On the other hand, with factor analysis, each input variable is viewed as a weighted combination of some underlying theoretical construct. In sum, the principal components method is just a transformation of the data to a new set of variables and is useful for prediction; factor analysis is a model for the data that seeks to explain the data.)

The weights $w_{ij}$ are computed by a mathematical method called eigenvalue analysis that is applied to the correlation matrix of the original data set. This is equivalent to standardizing each variable by subtracting its mean and dividing by its standard deviation. This standardization is important, lest results depend on the units of measurement. For example, if one of the variables is dollars, measuring that variable in cents or thousands of dollars will produce completely different sets of principal components. The eigenvalue analysis of the correlation matrix of k variables produces k eigenvalues, and for each eigenvalue there is an eigenvector of *k* elements. For example, the k elements of the first eigenvector are $w_{11}$, $w_{12}$, $w_{13}$ ..., $w_{1k}$.

To see how these weights work, let's continue with the toyprincomp.jmp data set. Click the red triangle next to Principal Components and click **Eigenvectors**. Added to the principal component report will be the eigenvectors as shown in Figure 6.5. Accordingly, the equations for Prin1 and Prin2 are

$$\text{Prin1} = 0.70711 \text{ x} + 0.70711 \text{ z}$$
$$\text{Prin2} = 0.70711 \text{ x} - 0.70711 \text{ z}$$

Now, again click the red triangle next to Principal Components, but this time click **Eigenvalues**. The eigenvalues are displayed as shown in Figure 6.5. Observe that Prin1 accounts for 78.76% of the variation in the data, because its eigenvalue (1.5752) is 0.7876 of the sum of all the eigenvalues (1.5752+0.4248 =2). The fact that the sum of the eigenvalues equals two is not a coincidence. Remember that the variables are standardized, so each variable has a variance equal to one. Therefore, the sum of the standardized variances equals the number of variables, in this case, two.

**Figure 6.5  Eigenvectors and Eigenvalues for the toyprincomp.jmp Data Set**

| Eigenvectors | | |
|---|---|---|
| | Prin1 | Prin2 |
| x | 0.70711 | 0.70711 |
| z | 0.70711 | -0.70711 |

| Eigenvalues | | | | | | | |
|---|---|---|---|---|---|---|---|
| Number | Eigenvalue | Percent | 20 40 60 80 | Cum Percent | ChiSquare | DF | Prob>ChiSq |
| 1 | 1.5752 | 78.760 | | 78.760 | 3.415 | 0.425 | 0.0206* |
| 2 | 0.4248 | 21.240 | | 100.000 | 0.000 | . | . |

Finally, we can verify that the principal components, Prin1 and Prin2, are uncorrelated by computing the correlation matrix for the principal components. We can generate the correlation matrix by selecting **Analyze→Multivariate Methods→Multivariate**; select **Prin1** and **Prin2** and click **OK**. The correlation matrix clearly shows that the correlation between Prin1 and Prin2 equals zero.

# Dimension Reduction

One of the main objectives of PCA is to reduce the information contained in all the original variables into a smaller set of components with a minimum loss of information. So, the question is how do we determine how many components should be considered. To answer this question, let's look at another data set, princomp.jmp, which has 12 independent variables, x1 through x12; a single dependent variable, y; and 100 observations. As we did earlier in this chapter with the toyprincomp.jmp data set, first generate the correlation matrix and scatterplot matrix:

Select **Analyze→Multivariate Methods→Multivariate**. For the Select Columns option in the Multivariate and Correlations dialog box, click **x1**, hold down the shift key, and click **x12** to select all twelve x variables. Click **Y,Columns**, and all the variables will appear in the white box to the right of **Y, Columns**. Click **OK**.

It is rather hard to make sense of this correlation matrix and the associated scatterplots, so let's try another way to see this information.

Click the red triangle next to Multivariate. Select **Pairwise Correlations**. Scroll to the bottom of the screen to see a table listing of the pairwise correlations. Right-click in the table and select **Sort by Column**. Select **Signif Prob** and check the box for Ascending. Click **OK**.

You should observe most of the correlations are rather low with four correlations in the 60% range and several variables simply uncorrelated. The correlations for about half the correlations are significant, indicated by an asterisk in the Signif Prob column, while about half the correlations are insignificant.

Next, let's perform the PCA on the variables x1 through x12:

Select **Analyze→Multivariate Methods→Principal Components**. In the Principal Components dialog box, click **x1**, hold down the shift key, and click **x12**. Then click **Y, Columns** and click **OK**. Click the red triangle and click **Eigenvalues**. Again, click the red triangle and click **Scree Plot**.

Figure 6.6 displays the Scree Plot and eigenvalues. Examining the Scree Plot, it appears that there may be an "elbow" at 2 or 3 or 4 principal components. When we examine the histogram of eigenvalues, it appears that the first three principal components account for about 60% of the variation in the data, and the first four principal components might account for about 68% of the variation. How many principal components should we select? There are three primary methods for choosing this number, and any one of them can be satisfactory. (No one is necessarily better than any of the others in all situations, so selecting the number of principal components is an art form.)

### Figure 6.6  Scree Plot and Eigenvalues for x1 to x12 from the princomp.jmp Data Set

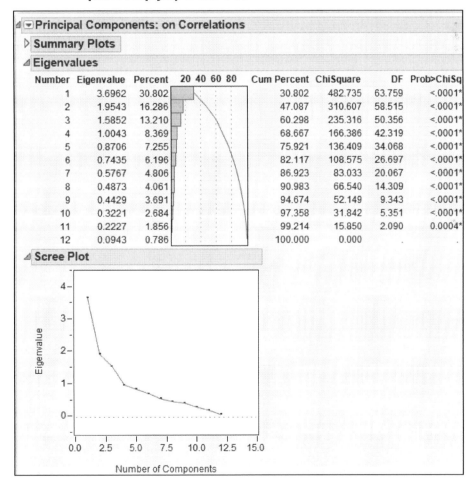

**METHOD ONE:** Look for an "elbow" in the scree plot of eigenvalues. This method will be covered in detail in Chapter 7, so our treatment here is brief. In our princomp.jmp data set, there is a clear elbow at 3, which suggests keeping the first three principal components.

**METHOD TWO:** How many eigenvalues are greater than one? Each principal component accounts for a proportion of variation related to its corresponding eigenvalue, and the sum of the eigenvalues equals the variance in the data set. Any principal component with an eigenvalue greater than one is contributing more than its share to the variance. Those principal components associated with eigenvalues that are less than one are not accounting for their share of the variation in the data. This suggests that we should retain principal components associated with eigenvalues greater than one. It is important not to be too strict with this rule. For example, if the fourth eigenvalue equals 1.01 and the fifth eigenvalue equals 0.99, it would be silly to use the first four principal components, because the fifth principal component makes practically the same contribution as the fourth principal component. This method is especially useful if there is no clear elbow in the scree plot. In our princomp.jmp example, we would choose 4 components.

**METHOD THREE:** Accounting for a specified proportion of the variation. This can be used two ways. First, the researcher can desire to account for at least, say, 70% or 80% of the variation, and retain enough principal components to achieve this goal. Second, the researcher can keep any principal component that accounts for more than, say, 5% or 10% of the total variation. For our princomp.jmp example, let us choose to explain at least 70% of the variation. We would then choose five principal components because four principal components only account for 68.667% of the variation.

Let's see what difference each of these methods makes with this data set. First, if we run a regression of y on all 12 x variables (*i.e.*, x1 to x12), the $R^2$ is 0.834434. The $R^2$ when we regress y on three, four, and five principal components is listed in Table 6.2. The difference in $R^2$ between three and four principal components is more than 4% (and the maximum $R^2$ is 100%), while the difference in $R^2$ between four and five is a negligible 0.001637. This suggests that four might be a good number of principal components to retain. The predictions that we get from keeping four or five principal components will be practically the same.

**Table 6.2 Regression Results**

|  | $R^2$ |
| --- | --- |
| y\|x1, x2, . . ., x12 | 0.834434 |
| y\| Princomp1 through Princomp3 | 0.723898 |
| y\| Princomp1 through Princomp4 | 0.766565 |
| y\| Princomp1 through Princomp5 | 0.768202 |

# Discovering Structure in the Data

In addition to dimension reduction, PCA can also be used to gain insight into the structure of the data set in two ways. First, the factor loadings can be used to plot the variables in the principal components space (this is the "Loading Plot"), and it is sometimes possible to see which variables are "close" to each other in the principal components space. Second, the principal component scores can be plotted for each observation (this is the "Score Plot"), and aberrant observations or small, unusual clusters might be noted. In this section, we consider both of these uses of PCA. We first use a data set of track and field records, in which the results are very clean and easy to interpret. We then use a real-world economic data set where the results are messier and more typical of the results obtained in practice.

The olymp88sas.jmp data set contains decathlon results for 34 contestants in the 1988 Summer Olympics in Seoul. In the decathlon, each contestant competes in 10 events, and an athlete's performance in each event makes a contribution to the final score; the contestant with the highest score wins.

First, as we have done before, generate the correlation matrix and scatterplot matrix: Select **Analyze→Multivariate Methods→Multivariate**. Select all the variables, click **Y, Columns**, and click **OK**. In the red triangle next to **Multivariate**, click **Pairwise Correlations**. Right-click in the table of pairwise correlations, select **Sort by Column** select **Signif Pro**, check the **Ascending Box**, and click **OK**. You should observe most of the correlations are rather significant with only a couple of correlations less than (absolute value, that is) 0.15.

Next, let's perform the PCA on this data set, but be sure to exclude "score". Why? Because "score" is a combination of all the other variables! After running the PCA, in the red triangle next to **Principal Components**, click **Eigenvalues** and **Scree Plot**. Figure 6.7 displays the PCA summary plots, eigenvalues, and Scree Plot for the data set. The bar chart of the eigenvalues indicates that two principal components dominate the data set, contributing 70% of the variability. The Score Plot shows that almost all the observations cluster near the origin of the space defined by the first two principal components, but there is a noticeable outlier on the far left. Clicking on this observation reveals that it is observation 34, the contestant with the worst score. Examine the score column further. First, realize that the observations were sorted from highest to lowest score. Next, notice that the difference between any two successive contestants' scores is usually less than 50, and occasionally one or two hundred points. Contestant 34, however, finished 1568 points behind contestant 33; contestant 34 certainly is an outlier.

Back to the PCA output and Figure 6.7, we can see in the Loading Plot how neatly the variables cluster into three groups. In the upper left quadrant are the races, in the upper right quadrant are the throwing events, and along the right side of the x-axis are the jumping events. With economic data, structures are rarely so well-defined and easily recognizable, as we shall see next.

**Figure 6.7  PCA Summary Plots, Eigenvalues, and Scree Plot for the olymp88sas.jmp Data Set**

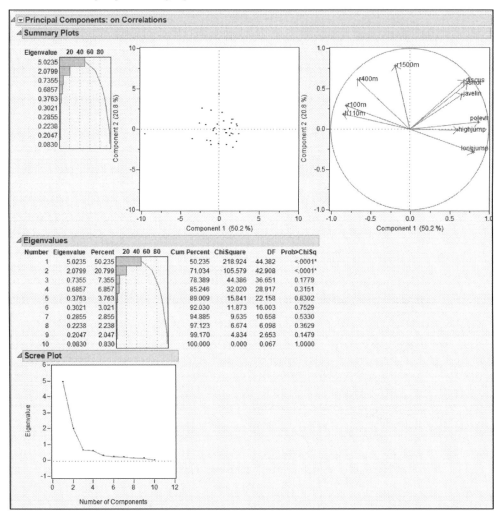

The Stategdp2008.jmp data set contains state gross domestic product (GDP) for each state and the District of Columbia for the year 2008 in billions of dollars. Each state's total GDP is broken down into twenty categories of state GDP, with total GDP given for each state in the second column and the total for the United States and each category given in the top row.

First, generate the correlation matrix and scatterplot matrix. The correlation matrix of the data set shows that the variables are highly correlated, with many correlations exceeding 0.9. This makes sense, because the data are measured in dollars and states with large economies.

For example, California, Texas, and New York tend to have large values for each of the categories, and small states tend to have small values for each of the categories. The fact that the data set is highly correlated suggests that the data set may have only one principal component. We encourage the reader to see this by running a PCA on the data set. Select all the numeric variables except total. Why? Try it both ways and see what happens.

The Scree plot shows that the first eigenvalue accounts for practically all of the variation; this confirms our earlier suspicion. Notice how the score plot shows almost all of the observations clustered at the origin, with a single observation on the x-axis near the value of 30. Click on that observation. What mistake have we made?

We included the "US" observation in the analysis, which we should not have done. To exclude this row from the analysis, on the JMP data table, in the list of row numbers in the first column (with the observation numbers), right-click on the first observation **1** and click **Exclude/Unexclude**. A red circle with a line through it should appear next to the row number. Now re-run the PCA, and the output will look like Figure 6.8. From the bar chart of the eigenvalues, we can see that the first eigenvalue dominates, but that a second eigenvalue might also be relevant. In the score plot, click on each of the three right-most data points to see that they represent NY, TX, and CA. This plot just shows us what we already know: that these are the three largest state economies. In the factor loadings plot, all the variables except mining are clustered together.

## Figure 6.8  PCA Summary Plots for the StateGDP2008.jmp Data Set

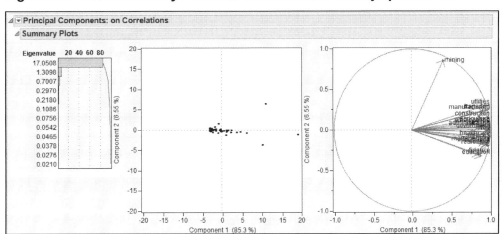

To obtain a more informative plot, it might be more useful to express the data in proportional form, to show what proportion each category constitutes of the state's total GDP. For example, the proportion of Alaska's (AK) state GDP by agriculture is $2.368/170.14 = 0.0139$. Let's now perform the analysis again using the StateGDP2008percent.jmp data set that has the relative proportions. Examine the correlation matrix of the data set; another way of producing the correlation matrix is to run the PCA. Then click the red triangle and click **Correlations**. Besides the main diagonal, which has to be all ones, there are no large correlations. We can expect to find a few important principal components, not just one. Indeed, as can be seen in Figure 6.9, the bar chart of the eigenvalues shows that the first two principal components account for barely 40% of the variation. In the lower left quadrant of the score plot, we find two observations that belong to Alaska and Wyoming, which are very negative on both the first and second principal components. Referring to the data table, we see that only two states have a proportion for mining that exceeds 20%; those two are Alaska and Wyoming. If we identify the other four observations that fall farthest from the origin in this quadrant, we find them to be LA, WV, NM, and OK, the states that derive a large proportion of GDP from mining activities. In the lower right corner, is a single observation DC that gets 32% of its GDP from government.

**Figure 6.9  PCA Summary Plots for the StateGDP2008percent.jmp Data Set**

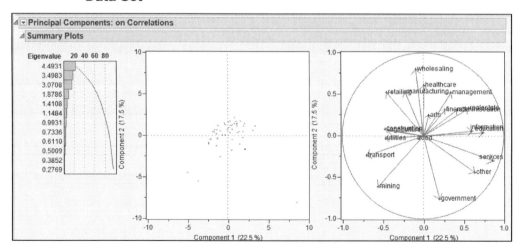

When there are only two or three important principal components, analyzing the loading plots involves looking at only 1 or perhaps 3 plots (in the former case, the first and second principal components; in the latter case, plots of first versus second, second versus third, and first versus third are necessary). For more than two important principal components, examining all the possible principal component plots becomes problematic. Nonetheless, for expository purposes let us look at the loadings plot. We see in the lower left quadrant that "transport" and "mining" both are negatively correlated with both the first and second principal components. In the lower right quadrant, "other" is negatively correlated with the second principal component but positively correlated with the first principal component, as are "services" and "government."

While this chapter has included two examples to illustrate the use of PCA for exploring the structure of a data set, it is important to keep in mind that the primary use of PCA in multivariate analysis and data mining is data reduction, that is, reducing several variables to a few. This concept will be explored further in the exercises.

---

# Exercises

1.  Use the PublicUtilities.jmp data set. Run a regression to predict **return** using all the other variables. Run a PCA and use only a few principal components to predict **return** (remember not to include **return** in the variables on which the PCA is conducted).

2.  Use the MassHousing.jmp data set. Run a regression to predict market value (**mvalue**) using all the other variables. Run a PCA and use only a few principal components to predict **mvalue** (remember not to include **mvalue** in the variables on which the PCA is conducted).

# Chapter 7

---

## Cluster Analysis

**Figure 7.1 A Framework for Multivariate Analysis**

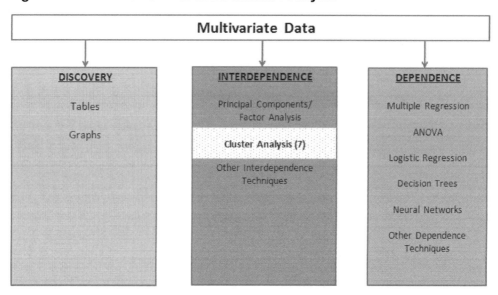

Cluster analysis is an exploratory multivariate technique designed to uncover natural groupings of the rows in a data set. If data are two-dimensional, it can be very easy to find groups that exist in the data; a scatterplot will suffice. When data have three, four, or more dimensions, how to find groups is not immediately obvious. As shown in Figure 7.1, cluster analysis is a technique where no dependence in any of the variables is required. The object of cluster analysis is to divide the data set into groups, where the observations within each group are relatively homogeneous, yet the groups are unlike each other.

A standard clustering application in the credit card industry is to segment its customers into groups based on the number of purchases made, whether balances are paid off every month, where the purchases are made, etc. In the cell-phone industry, clustering is used to identify customers who are likely to switch carriers. Grocery stores with loyalty-card programs cluster their customers based on the number, frequency, and types of purchases. After customers are segmented, advertising can be targeted. For example, there is no point in sending coupons for baby food to all the store's customers, but sending the coupons to customers who have recently purchased diapers might be profitable. Indeed, coupons for premium baby food can be sent to customers who have recently purchased filet mignon, and coupons for discount baby food can be sent to customers who have recently purchased hamburger.

A recent cluster analysis of 1000 credit card customers from a commercial bank in Shanghai identified three clusters (*i.e.*, market segments). The analysis described them as follows (Ying and Yuanuan, 2010):

1. First Class: married, between 30 and 45 years old, above average salaries with long credit histories and good credit records.

2. Second Class: single, under 30, fashionable, no savings and good credit records, and tend to carry credit card balances over to the next month.

3. Third Class: single or married, 30-45, average to below-average incomes with good credit records.

Each cluster was then analyzed in terms of contribution to the company's profitability and associated risk of default, and an appropriate marketing strategy was designed for each group.

The book *Scoring Points* by Humby, Hunt, and Phillips (2007) tells the story of how the company Tesco used clustering and other data mining techniques to rise to prominence in the British grocery industry. After much analysis, Tesco determined that its customers could be grouped into 14 clusters. For example, one cluster always purchased the cheapest commodity in any category; Tesco named them "shoppers on a budget." Another cluster contained customers with high incomes but little free time, and they purchased high-end, ready-to-eat food.

Thirteen of the groups had interpretable clusters, but the buying patterns of the fourteenth group, which purchased large amounts of microwavable food, didn't make sense initially. After much investigation, it was determined that this fourteenth group actually comprised two groups that the clustering algorithm had assigned to a single cluster: young people in group houses who did not know how to cook (so bought lots of microwavable food) and families with traditional tastes who just happened to like microwavable food. The two otherwise disparate groups had been clustered together based on their propensity to purchase microwavable food. But the single purchase pattern had been motivated by the unmet needs of two different sets of customers.

Two morals are evident. First, clustering is not a purely data-driven exercise. It requires careful statistical analysis and interpretation by an industry business expert to produce good clusters. Second, many iterations may be needed to achieve the goal of producing good clusters, and some of these iterations may require field-testing. In this chapter, we present two important clustering algorithms: hierarchical clustering and k-means clustering. Clustering is not a purely statistical exercise, and a good use of the method requires knowledge of statistics and of the characteristics of the business problem and the industry.

# Hierarchical Clustering

The specific form of hierarchical clustering used in JMP is called an agglomerative algorithm. At the start of the algorithm, each observation is considered as its own cluster. The distance between each cluster and all other clusters is computed, and the nearest clusters are merged. If there are n observations, this process is repeated n-1 times until there is only one large cluster. Visually, this process is represented by a tree-like figure called a dendrogram. Inspecting the dendrogram allows the user to make a judicious choice about the number of clusters to use. Sometimes an analyst will want to perform k-means clustering–which requires that the number of clusters be specified–but have no idea of how many clusters are in the data. Perhaps the analysts want to name the clusters based on a k-means clustering. In such a situation, one remedy is to perform hierarchical clustering first in order to determine the number of clusters. Then k-means clustering is performed.

Except at the very first step when each observation is its own cluster, clusters are not individual observations/subjects/customers, but collections of observations/subjects/customers. There are many ways to calculate the distance between two clusters. Figure 7.2 shows four methods of measuring the distance between two clusters. An additional method, which is not amenable to graphical depiction, is Ward's method, which is based on minimizing the information loss that occurs when two clusters are joined.

The various methods of measuring distance tend to produce different types of clusters. Average linkage is biased toward producing clusters with the same variance. Single linkage imposes no constraint on the shape of clusters, and makes it easy to combine two clumps of observations that other methods might leave separate. Hence, single linkage has a tendency toward what is called "chaining," and can produce long and irregularly shaped clusters. Complete linkage is biased toward producing clusters with similar diameters. The centroid method is more robust to outliers than other methods, while Ward's method is biased toward producing clusters with the same number of observations.

**Figure 7.2 Ways to Measure Distance between Two Clusters**

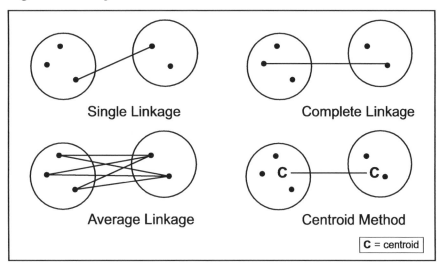

Figure 7.3 shows how single linkage can lead to combining clusters that produce long strings, while complete linkage can keep them separate. The horizontal lines indicate the distance computed by each method. By the single linkage method, the left and right groups are very close together and, hence, are combined into a single cluster. In contrast, by the complete linkage method, the two groups are far apart and are kept as separate clusters.

**Figure 7.3 The Different Effects of Single and Complete Linkage**

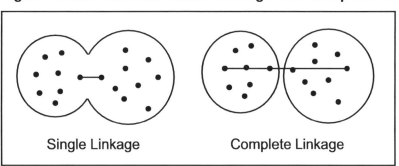

To get a feeling for how hierarchical clustering works, consider the toy data set in Table 7.1 and in the file Toy-cluster.jmp.

### Table 7.1  Toy Data Set for Illustrating Hierarchical Clustering

| X | Y | Label |
|---|---|-------|
| 1 | 1 | A |
| 1 | 0 | B |
| 2 | 2 | C |
| 3 | 2 | D |
| 4 | 3 | E |
| 5 | 4 | F |
| 6 | 5 | G |
| 6 | 6 | H |
| 2 | 5 | I |
| 3 | 4 | J |

To perform hierarchical clustering with complete linkage on the Toy-cluster.jmp data table:

1. Select **Analyze→Multivariate Methods→Cluster**.

2. From the Fit Model dialog box, select both **X** and **Y** and click **Y, Columns**. Also, click **Label**. Since the units of measurement can affect the results, make sure that Standardize Data is checked. Under Options, select Hierarchical and Ward (see Figure 7.4).

3. Click **OK**.

## Figure 7.4 The Clustering Dialog Box

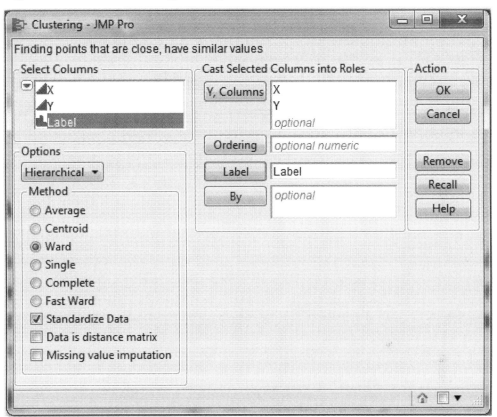

The clustering output includes a dendrogram similar to Figure 7.5. Click the red triangle to the left of Hierarchical Clustering and click **Color Clusters** and **Mark Clusters**. Next, notice the two diamonds, one at the top and one at the bottom of the dendrogram. Click one of them, and the number 2 should appear. The 2 is the current number of clusters, and you can see how the observations are broken out. Click one of the diamonds again and drag it all the way to the left where you have each observation, alone, in its individual cluster. Now arrange the JMP windows so that you can see both the data table and the dendrogram. Slowly drag the diamond to the right and watch how the clusters change as well as the corresponding colors and symbols. Figure 7.6 is a scatterplot of the 10 observations in the Toy data set.

**Figure 7.5 Dendogram of the Toy Data Set**

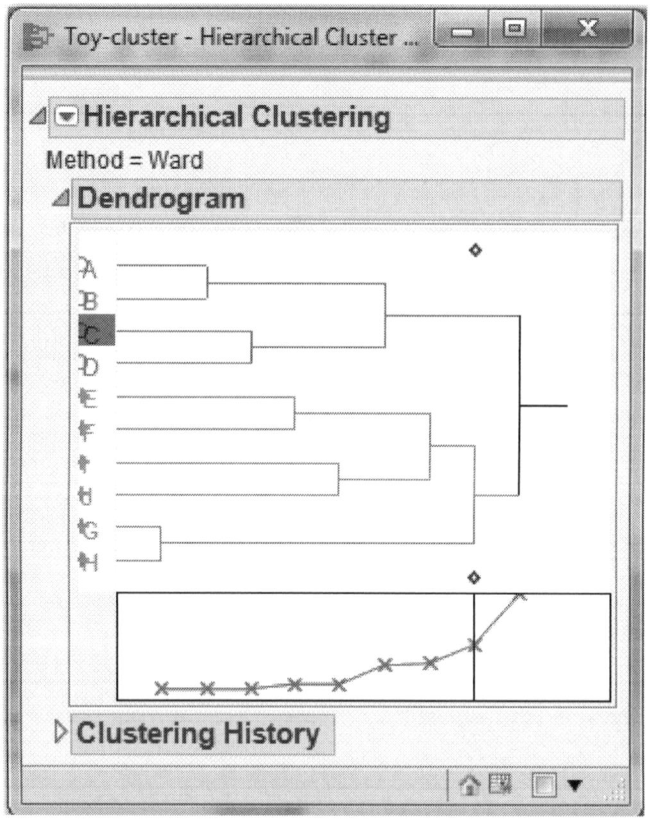

**Figure 7.6  Scatterplot of the Toy Data Set**

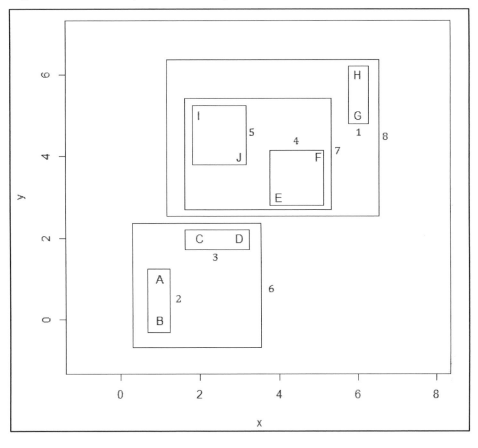

Looking at the scatterplot in Figure 7.6, the clusters are numbered in the order in which they were created, following the steps below. Click the arrow to the left of **Clustering History**. When you move the diamond from left to right, the dendrogram and the clustering history can be read as follows:

**STEP 1:** H and G are paired to form a cluster.

**STEP 2:** A and B are paired to form a cluster.

**STEP 3:** C and D are paired to form a cluster.

**STEP 4:** E and F are paired to form a cluster.

**STEP 5:** I and J are paired to form a cluster.

**STEP 6**: The AB and CD clusters are joined.

**STEP 7:** The EF and IJ clusters are combined.

**STEP 8:** HG is added to the EFIJ cluster.

**STEP 9:** The ABCD and EFIJGH clusters are combined.

The clustering process is more difficult to imagine in higher dimensions, but the idea is the same.

The fundamental question when applying any clustering technique, and hierarchical clustering is no exception, is "What is the best number of clusters?" There is no target variable and no a priori knowledge of which customer belongs to which group, so there is no gold standard by which we can determine whether any particular classification of observations into groups is successful or not. Sometimes it is enough simply to recognize that a particular classification has produced groups that are interesting or useful in some sense or another. In the absence of such compelling anecdotal justifications, one common method is to use the scree plot beneath the dendrogram to gain some insight into what the number of cluster might be. The ordinate (y-axis) is the distance that was bridged in order to join the clusters, and a natural break in this distance produces a change in the slope of the line (often described as an "elbow") that suggests an appropriate number of clusters. At the bottom of the box in Figure 7.5 is a small scree plot with a line of x's. Each x represents the level of clustering at each step. Since there are 10 observations, there are 10 x's. The elbow appears to be around 2 or 3 clusters, so 2 or 3 clusters might be appropriate for these data.

Let us now apply these principles to a data set, Thompson's "1975 Public Utility Data Set" (Johnson and Wichern, 2002). The data set is small enough so that all the observations can be comprehended and rich enough so that it provides useful results; it is a favorite for exhibiting clustering methods. Table 7.2 shows the Public Utility Data found in PublicUtilities.jmp, which has eight numeric variables that will be used for clustering: coverage, the fixed-charge coverage ratio; return, the rate of return on capital; cost, cost per kilowatt hour; load, the annual load factor; peak, the peak kilowatt hour growth; sales, in kilowatt hours per year; nuclear, the percent of power generation by nuclear plants; and fuel, the total fuel costs. "Company" is a label for each observation; it indicates the company name.

## Table 7.2 Thompson's 1975 Public Utility Data Set

| Coverage | Return | Cost | Load | Peak | Sales | Nuclear | Fuel | Company |
|---|---|---|---|---|---|---|---|---|
| 1.06 | 9.20 | 151 | 54.4 | 1.6 | 9077 | 0 | 0.63 | Arizona Public |
| 0.89 | 10.3 | 202 | 57.9 | 2.2 | 5088 | 25.3 | 1.555 | Boston Edison |
| 1.43 | 15.4 | 113 | 53 | 3.4 | 9212 | 0 | 1.058 | Central Louisiana |
| 1.02 | 11.2 | 168 | 56 | 0.3 | 6423 | 34.3 | 0.7 | Commonwealth Edison |
| 1.49 | 8.8 | 192 | 51.2 | 1 | 3300 | 15.6 | 2.044 | Consolidated Edison |
| 1.32 | 13.5 | 111 | 60 | -2.2 | 11127 | 22.5 | 1.241 | Florida Power & Light |
| 1.22 | 12.2 | 175 | 67.6 | 2.2 | 7642 | 0 | 1.652 | Hawaiian Electric |
| 1.1 | 9.2 | 245 | 57 | 3.3 | 13082 | 0 | 0.309 | Idaho Power |
| 1.34 | 13 | 168 | 60.4 | 7.2 | 8406 | 0 | 0.862 | Kentucky Utilities |
| 1.12 | 12.4 | 197 | 53 | 2.7 | 6455 | 39.2 | 0.623 | Madison Gas |
| 0.75 | 7.5 | 173 | 51.5 | 6.5 | 17441 | 0 | 0.768 | Nevada Power |
| 1.13 | 10.9 | 178 | 62 | 3.7 | 6154 | 0 | 1.897 | New England Electric |
| 1.15 | 12.7 | 199 | 53.7 | 6.4 | 7179 | 50.2 | 0.527 | Northern States Power |
| 1.09 | 12 | 96 | 49.8 | 1.4 | 9673 | 0 | 0.588 | Oklahoma Gas |
| 0.96 | 7.6 | 164 | 62.2 | -0.1 | 6468 | 0.9 | 1.4 | Pacific Gas |
| 1.16 | 9.9 | 252 | 56 | 9.2 | 15991 | 0 | 0.62 | Puget Sound Power |
| 0.76 | 6.4 | 136 | 61.9 | 9 | 5714 | 8.3 | 1.92 | San Diego Gas |
| 1.05 | 12.6 | 150 | 56.7 | 2.7 | 10140 | 0 | 1.108 | The Southern Co. |
| 1.16 | 11.7 | 104 | 54 | -2.1 | 13507 | 0 | 0.636 | Texas Utilities |
| 1.2 | 11.8 | 148 | 59.9 | 3.5 | 7287 | 41.1 | 0.702 | Wisconsin Electric |
| 1.04 | 8.6 | 204 | 61 | 3.5 | 6650 | 0 | 2.116 | United Illuminating |
| 1.07 | 9.3 | 174 | 54.3 | 5.9 | 10093 | 26.6 | 1.306 | Virginia Electric |

To perform hierarchical clustering on the Public Utility Data Set, open the
PublicUtilities.jmp file:

1. Select **Analyze→Multivariate Methods→Cluster**.

2. From the Clustering dialog box, select Coverage, Return, Cost, Load, Peak,
   Sales, Nuclear, and Fuel and click **Y, Columns**. Also, select Company and click
   Label. Since the units of measurement can affect the results, make sure that
   Standardize Data is checked. Under Options, choose Hierarchical and Ward.

3. Click **OK**.

4. In the Hierarchical Clustering output, click the red triangle and click Color
   Clusters and Mark Clusters.

**Figure 7.7 Hierarchical Clustering of the Public Utility Data Set**

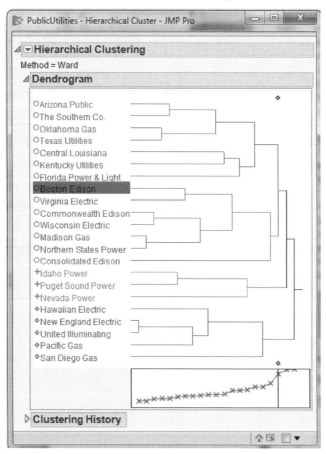

The results are shown in Figure 7.7. The vertical line in the box goes through the third (from the right) x, which indicates that three clusters might be a good choice. The first cluster is Arizona Public through Consolidated Edison (14 firms); the second cluster is Idaho, Puget Sound, and Nevada (three firms); and the third cluster is Hawaiian through San Diego (five firms). We can move the vertical line and look at the different clusters. It appears that clusters of size 3, 4, 5 or 6 would be "best." For now, let us suppose the Ward's method with 5 clusters suffices. Move the vertical line so that it produces five clusters. Let's produce a report/profile of this "5-cluster solution. " In practice, we may well produce similar reports for the 3-, 4-, and 6-cluster solutions also, for review by a subject-matter expert who could assist us in determining the appropriate number of clusters.

To create a column in the worksheet that indicates each firm's cluster, click the red triangle and click **Save Clusters**. To help us interpret the clusters, we will produce a table of means for each cluster:

1.  In the data table, select **Tables→Summary**. Select Coverage, Return, Cost, Load, Peak, Sales, Nuclear, and Fuel, click the Statistics box, and select **Mean**. Next, select the variable **Cluster** and click **Group**.

2.  Click **OK**. The output contains far too many decimal places. To remedy this, select all the columns. Then select **Cols→Column Info**. Under Format, select Fixed Dec and then change Dec from 0 to 2 for each variable. Click **OK**.

The output presented is shown in Table 7.3, shown below.

## Table 7.3  Cluster Means for Five Clusters

| Cluster | N Rows | Mean(Cov erage) | Mean(Return) | Mean(Cost) | Mean(Load) | Mean(Peak) | Mean(Sales) | Mean(Nuclear) | Mean(Fuel) |
|---|---|---|---|---|---|---|---|---|---|
| 1 | 7 | 1.21 | 12.49 | 127.57 | 55.47 | 1.71 | 10163.14 | 3.21 | 0.87 |
| 2 | 6 | 1.08 | 11.28 | 181.33 | 55.80 | 3.50 | 7087.50 | 36.12 | 0.90 |
| 3 | 1 | 1.49 | 8.80 | 192.00 | 51.20 | 1.00 | 3300.00 | 15.60 | 2.04 |
| 4 | 3 | 1.00 | 8.87 | 223.33 | 54.83 | 6.33 | 15504.67 | 0.00 | 0.57 |
| 5 | 5 | 1.02 | 9.14 | 171.40 | 62.94 | 3.66 | 6525.60 | 1.84 | 1.80 |

To understand a data set via clustering, it can be useful to try to identify the clusters by looking at summary statistics. In the present case, we can readily identify all the clusters: Cluster 1 is highest return; Cluster 2 is highest nuclear; Cluster 3 is the singleton; Cluster 4 is highest cost; and Cluster 5 is highest load. Here we have been able to characterize each cluster by appealing to a single variable. But often it will be necessary to use two or more variables to characterize a cluster. Additionally, we recognize that we have clustered based on standardized variables and identified them using the original variables. Sometimes it may be necessary to refer to the summary statistics of the standardized variables to be able to identify the clusters.

# Using Clusters in Regression

In Problem 6 of the Chapter 4 exercises, the problem was to run a regression on the PublicUtilities data using sales as a dependent variable and Coverage, Return, Cost, Load, Peak, Sales, Nuclear, and Fuel as independent variables. The model had a $R^2$ of 59.3%. Not too high, and the adjusted $R^2$ is only 38.9%. What if we add the nominal variable Clusters to the regression model? The model improves significantly with a $R^2$ of 88.3%, and the adjusted $R^2$ is only 75.5%. But now we have 12 parameters to estimate (8 numeric variables and 4 dummy variables) and only 22 observations. This high $R^2$ is perhaps a case of overfitting. Overfitting occurs when the analyst includes too many parameters and therefore fits the random noise rather than the underlying structure; this concept is discussed in more detail in Chapter 10. Nevertheless, if you run the multiple regression with only the Cluster variable (thus, 4 dummy variables), the $R^2$ is 83.6% and the adjusted $R^2$ is 79.8%. We have dropped the 8 numeric variables from the regression, and the $R^2$ only dropped from 88.3% to 83.6%. This strongly suggests that the clusters contain much of the information that is in the 8 numeric variables. Using a well-chosen cluster variable in regression can prove to be very useful.

# K-means Clustering

While hierarchical clustering allows examination of several clustering solutions in one dendrogram, it has two significant drawbacks when applied to large data sets. First, it is computationally intensive and can have long run times. Second, the dendrogram can become large and unintelligible when the number of observations is even moderately large.

One of the oldest and most popular methods for finding groups in multivariate data is the *k*-means clustering algorithm, which has five steps:

1.  Choose k, the number of clusters.

2.  Guess at the multivariate means of the *k* clusters. If there are ten variables, each cluster will be associated with ten means, one for each variable. Very often this collection of means is called a centroid. JMP will perform this guessing with the assistance of a random number generator, which is a mathematical formula that produces random numbers upon demand. It is much easier for the computer to create these guesses than for the user to create them by hand.

3.  For each observation, calculate the distance from that observation to each of the *k* centroids, and assign that observation to the closest cluster (*i.e.*, the closet centroid).

4. After all the observations have been assigned to one and only one cluster, calculate the new centroid for each cluster using the observations that have been assigned to that cluster. The cluster centroids "drift" toward areas of high density, where there are many observations.

5. If the new centroids are very different from the old centroids, the centroids have drifted. So return to Step 3. If the new centroids and the old centroids are the same so that additional iterations will not change the centroids, then the algorithm terminates.

The effect of the k-means algorithm is to minimize the differences within each group, and to maximize the differences between groups, as shown in Figure 7.8.

**Figure 7.8 The Function of *k*-means Clustering**

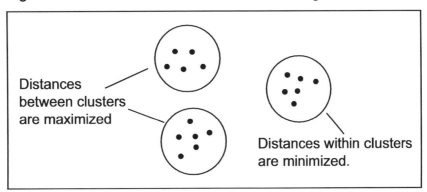

A primary advantage of the *k*-means clustering algorithm is that it has low complexity. By this we mean that its execution time is proportional to the number of observations, so it can be applied to large data sets. By contrast, when you use an algorithm with high complexity, doubling the size of the data set may increase the execution time by a factor of four or more. Hence, algorithms with high complexity are not desirable for use with large data sets.

A primary disadvantage of the algorithm is that its result, the final determination of which observations belong to which cluster, can depend on the initial guess (as in Step 1 above). As a consequence, if group stability is an important consideration for the problem at hand, it is sometimes advisable to run the algorithm more than once to make sure that the groups do not change appreciably from one run to another. All academics advise multiple runs with different starting centroids, but practitioners rarely do this, though they should. However, they have to be aware of two things: (1) comparing solutions from different starting points can be tedious, and the analyst will often end up observing, "this observation was put in cluster 1 on the old output, and now it is put in cluster 5 on the new output—I think"; and (2) doing this will add to the project time.

If the data were completely dependent, there would be a single cluster. If the data were completely independent, there would be as many clusters as there are observations. Almost always the true number of clusters is somewhere between these extremes, and the analyst's job is to find that number. If there is, in truth, only one cluster, but k is chosen to be five, the algorithm will impose five clusters on a data set that consists of only a single cluster, and the results will be unstable. Every time the algorithm is run, a completely different set of clusters will be found. Similarly, if there are really 20 clusters and k is chosen for the number of clusters, the results will again be unstable.

After the algorithm has successfully terminated, each observation is assigned to a cluster, each cluster has a centroid, and each observation has a distance from the centroid. The distance of each observation from the centroid of its cluster is calculated. Square them, and sum them all to obtain the sum of squared errors (SSE) for that cluster solution. When this quantity is computed for various numbers of clusters, it is traditional to plot the SSE and to choose the number of clusters that minimizes the sum of squared errors. For example, in Figure 7.9, the value of k that minimizes the sum of squared errors is four.

**Figure 7.9  U-shaped SSE plot for Choosing the Number of Clusters**

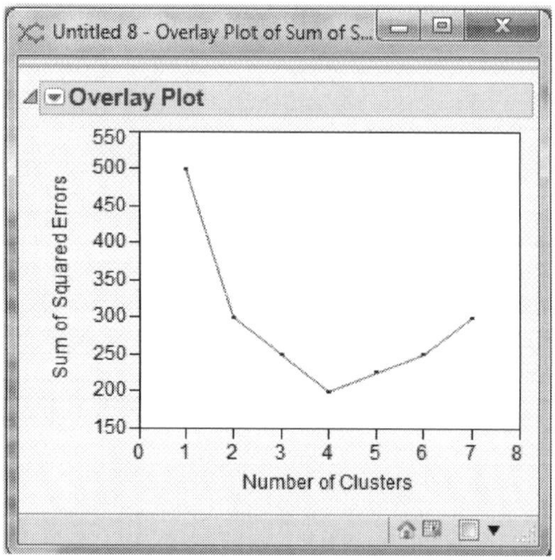

There is no automated procedure for producing a graph such as that in Figure 7.9. Every time the clustering algorithm is run, the user has to write down the sum of squared errors, perhaps entering both the sum of squared errors and the number of clusters in appropriately labeled columns in a JMP data table. Then you select **Graph→Overlay Plot**, and select the sum of squared errors as **Y** and the number of clusters as **X**. Click **OK** to produce a plot of points. To draw a line through the points, in the overlay plot window, click the red arrow and select **Connect Thru Missing**.

As shown in Figure 7.10, it is not always the case that the SSE will take on a U-shape for increasing number of clusters. Sometimes the SSE decreases continuously as the number of clusters increases. In such a situation, choosing the number of clusters that minimizes the SSE would produce an inordinately large number of clusters. In this situation, the graph of SSE versus the number of clusters is called a scree plot. Often there is a natural break where the distance jumps up suddenly. These breaks suggest natural cutting points to determine the number of clusters. The "best" number of clusters is typically chosen at or near this "elbow" of the curve. The elbow suggests which clusters should be profiled and reviewed with the subject-matter expert. Based on Figure 7.10, the number of clusters would probably be 3, but 2 or 4 would also be possibilities. Choosing the "best" number of clusters is as much an art form as it is a science. Sometimes a particular number of clusters produces a particularly interesting or useful result; in such a case, SSE can probably be ignored.

## Figure 7.10  A Scree Plot for Choosing the Number of Clusters

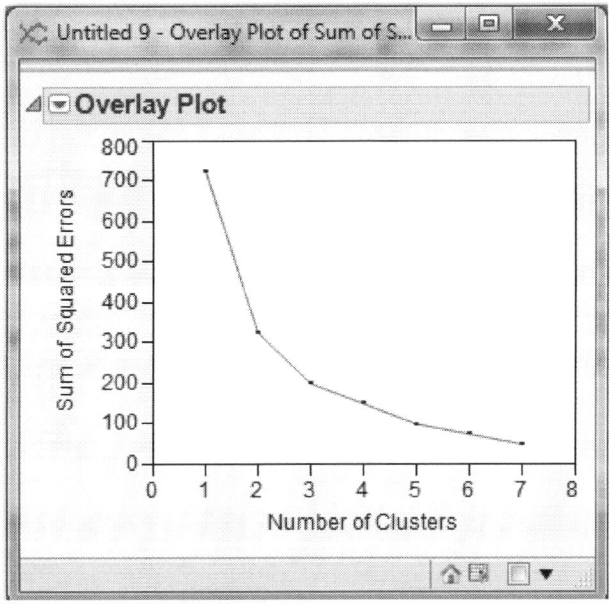

The difference between two and three or three and four clusters (indeed between any pair of partitions of the data set) is not always statistically obvious. The fact is that it can be difficult to decide on the number of clusters. Yet this is a very important decision, because the proper number of clusters can be of great business importance. In Figure 7.11, it is hard to say whether the data set has two, three, or four clusters. A pragmatic approach is necessary in this situation: choose the number of clusters that produces useful clusters.

Once k has been chosen and clusters have been determined, the centroids (multivariate) means of the clusters can be used to give descriptions (and descriptive names) to each cluster/segment. For example, a credit card company might observe that in one cluster, the customers charge large balances and pay them off every month. This cluster might be called "transactors." Another cluster may occasionally make a large purchase and then pay the balance down over time; they might be "convenience users." A third cluster would be customers who always have a large balance and roll it over every month, incurring high interest fees; they could be called "revolvers" since they use the credit card as a form of revolving debt. There would, of course, be other clusters, and perhaps not all of them would have identifiable characteristics that lead to names. The idea, however, should be clear.

**Figure 7.11  How Many Clusters in the Data Set?**

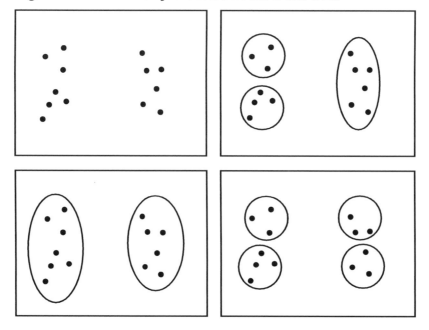

To perform k-means clustering on the PublicUtilities data set open JMP and load PublicUtilities.jmp:

1. Select **Analyze→Multivariate Methods→Cluster**.

2. From the Clustering dialog box, select Coverage, Return, Cost, Load, Peak, Sales, Nuclear, and Fuel (click Coverage, hold down the **Shift** key, and click Fuel) and click **Y, Columns**. Also, select Company and click **Label**.

3. Click the Options drop-down menu and change **Hierarchical** to **Kmeans**. Now, the k-means algorithm is sensitive to the units in which the variables are measured. If we have three variables (length weight, and value), we will get one set of clusters if the units of measurement are inches, pounds, and dollars, and (probably) a radically different set of clusters if the units of measurement are feet, ounces, and cents. To avoid this problem, make sure the box for "**Columns Scaled Individually**" is checked. The box for "**Johnson Transform**" is another approach to dealing with this problem that will balance skewed variables or bring outliers closer to the rest of the data; do not check this box.

4. Click **OK**.

A new pop-up box similar to Figure 7.12 appears. The Methodmenu indicates K-Means Clustering and the number of clusters is 3. But we will change the number of clusters shortly. (Normal Mixtures, Robust Normal Mixtures, and Self Organizing Map are other clustering methods about which you can read in the online user-guide and with which you may wish to experiment. To access this document, from the JMP Home Window, select **Help→Books→Modeling and Multivariate Methods** and consult Chapter 20 on "Clustering.")

Suppose we wanted more than 3 clusters. Then you would change the number of clusters to 5. The interested reader can consult the online help files to learn about "Single Step," "Use within-cluster std deviations," and "Shift distances using sampling rates." But do not check these boxes until you have read about these methods. The JMP input box should look like Figure 7.12. Click **Go** to perform *k*-means clustering. The output is shown in Figure 7.13.

### Figure 7.12  KMeans Dialog Box

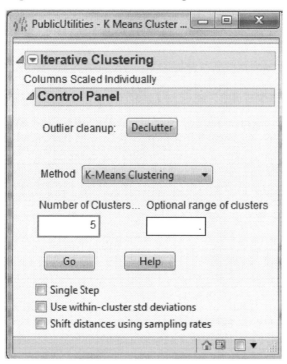

## Figure 7.13 Output of k-means Clustering with Five Clusters

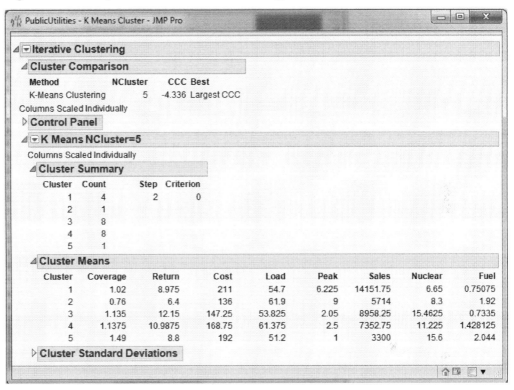

Under Cluster Summary in Figure 7.13, we can see that clusters 2 and 5 have a single observation; perhaps 5 clusters with the *k*-means is too coarse of a breakdown. So, let's rerun the k-means analysis with 3 clusters:

1. In the **K Means Cluster** dialog box, click the triangle next to the K Means NCluster=5 panel to collapse those results.

2. Click the triangle next to **Control Panel** to expand it, and change Number of Clusters to 3.

3. Click **Go**. The output will look like Figure 7.14.

**Figure 7.14 Output of k-means Clustering with Three Clusters**

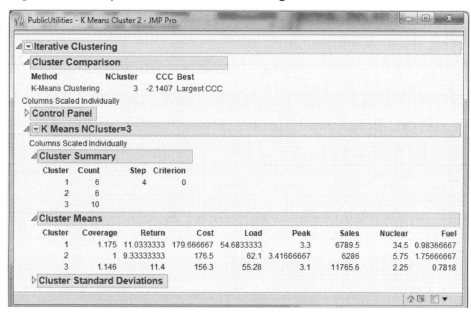

Optionally, click on the red triangle next to **K Means NCluster=3** and click **Parallel Coord Plots**. This command sequence creates plots of the variable means within each cluster, as shown in Figure 7.15. Parallel coordinate plots may be helpful in the interpretation of the clusters. This is a fast way to create a "profile" of the clusters, and can be of great use to a subject-matter expert. From Figure 7.15, we can see that Cluster 1 is the "high nuclear" group; this can be confirmed by referring to the actual data in Figure 7.14. For an excellent and brief introduction to parallel coordinate plots, see Few (2006).

**Figure 7.15 Parallel Coordinate Plots for k=3 Clusters**

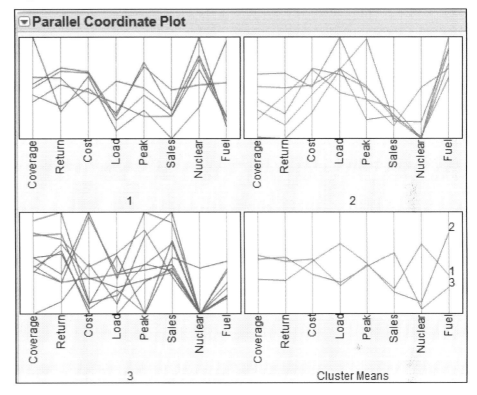

Under **Cluster Summary**, the program identifies three clusters with six, six, and ten companies in each. Using the numbers under **Cluster Means**, we can identify Cluster 1 as the "high nuclear" cluster and Cluster 3 as the "high sales" cluster. Nothing really stands out about Cluster 2, but perhaps it might not be too much of a stretch to refer to it as the "high load" or "high fuel" cluster. Since we are going to try other values of k before leaving this screen, we should calculate the SSE for this partition of the data set:

4. Click the red triangle next to **K Means NCluster = 3** and click **Save Clusters**. Two new columns will appear on the right of the data table. **Cluster** indicates the cluster to which each company has been assigned. **Distance** gives the distance of each observation from the centroid of its cluster. We need to square all these distances and sum them to calculate the SSE.

5. From the top menu select **Cols→New Column**. For **Column Name**, enter **Distance Squared**. Go to the bottom of the dialog box, click **Column Properties** and select **Formula**. The Formula Editor appears, and on the left select the variable Distance. You may have to scroll down to find it. Distance will appear in

the formula box. In the middle of the dialog box, are several operator buttons. Click the operator button $x^y$ and then type the number **2**, if necessary (the number 2 might already be there). See Figure 7.16.

**Figure 7.16  Formula Editor for Creating Distance Squared**

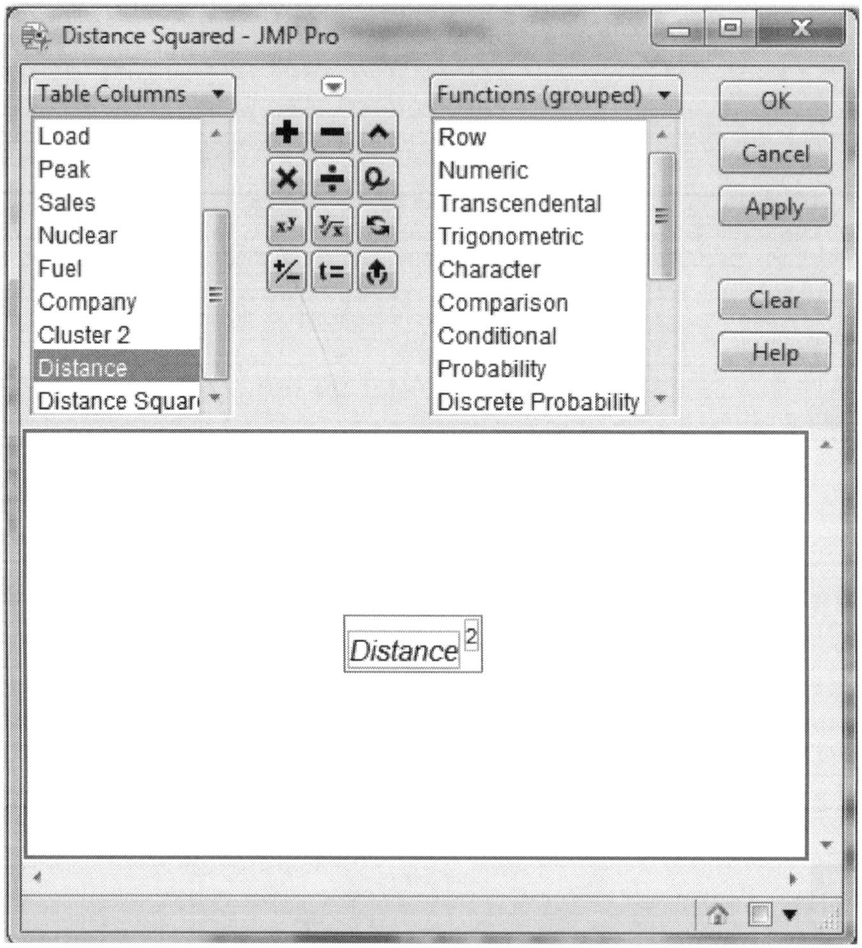

6. Click **OK**. The original **New Column** dialog box is still present, so click **OK** in it. The variable Distance Squared has been created, and we now need its sum.

7. On the data table menu, select **Analyze→Distribution**. Click Distance Squared and then click **Y, Columns**.

8. In the **Distribution** window, click the red arrow next to **Summary Statistics** (which is beneath Quantiles, which, in turn, is beneath the histogram/boxplot). Click **Customize Summary Statistics** and check the box for Sum. Click **OK**. Sum is now displayed along with the other summary statistics, and is seen to be 2075.1125. (Alternatively, on the top menu, click **Tables→Summary**. The Summary dialog box appears. From the list of Select Columns, click Distance Squared. Click the drop-down arrow next to Statistics and click Sum. Click **OK**.)

With the same number of clusters, the members of clusters will most likely differ from the Hierarchical Clustering and the k-means clustering. As a result, this procedure of producing the SSE using the k-means clustering should be performed for k = 4, 5, and 6. This iterative process can be facilitated by clicking the right arrow next to the **Control Panel** in the **Iterative Clustering** output and entering 4 in the Number of Clusters field and 6 in the Optional range of clusters field. When both of these boxes have numbers, the former is really the lower limit of the number of clusters desired, and the latter is really the upper limit of the number of clusters desired. Be careful not to check the boxes on the second screen if you want to reproduce the numbers below. The SSE for the different cluster sizes are listed in Table 7.4. By inspection, we see that a plot of k versus SSE would not be a scree plot, but instead would be u-shaped with a minimum for k = 5.

**Table 7.4  SSE for Various Values of *k***

| K | 3 | 4 | 5 | 6 |
|---|---|---|---|---|
| SSE | 2075.11 | 2105.49 | 1743.93 | 1900.59 |

Back to the k-means with 5 clusters, as we stated before, there are two singletons, which can be considered outliers that belong in no cluster. The clusters are given in Table 7.5.

Since clusters 2 and 5 are singletons that may be considered outliers, we need only try to interpret cluster 1, 3, and 4. The cluster means are given in Figure 7.17. Cluster 1 could be highest cost, and cluster 3 could be highest nuclear. Cluster 4 does not immediately stand out for any one variable, but might accurately be described as high (but not highest) load and high fuel.

As might be deduced from this simple example and examining it using hierarchical clustering and *k*-means clustering, if the data set is large with numerous variables, the prospect of searching for k is daunting. But that is the analyst's task: to find a good number for k.

### Table 7.5  Five Clusters for the Public Utility Data

| | |
|---|---|
| Cluster 1 (4) | Idaho Power, Nevada Power, Puget Sound Electric, Virginia Electric |
| Cluster 2 (1) | San Diego Gas |
| Cluster 3 (8) | Arizona Public, Central Louisiana, Commonwealth Edison, Madison Gas, Northern States Power, Oklahoma Gas, The Southern Co., Texas Utilities |
| Cluster 4 (8) | Boston Edison, Florida Power & Light, Hawaiian Electric, Kentucky Utilities, New England Electric, Pacific Gas, Wisconsin Electric, United Illuminating |
| Cluster 5 (1) | Consolidated Edison |

### Figure 7.17  Cluster Means for Five Clusters Using *k*-means

⊿ Cluster Means

| Cluster | Coverage | Return | Cost | Load | Peak | Sales | Nuclear | Fuel |
|---|---|---|---|---|---|---|---|---|
| 1 | 1.02 | 8.975 | 211 | 54.7 | 6.225 | 14151.75 | 6.65 | 0.75075 |
| 2 | 0.76 | 6.4 | 136 | 61.9 | 9 | 5714 | 8.3 | 1.92 |
| 3 | 1.135 | 12.15 | 147.25 | 53.825 | 2.05 | 8958.25 | 15.4625 | 0.7335 |
| 4 | 1.1375 | 10.9875 | 168.75 | 61.375 | 2.5 | 7352.75 | 11.225 | 1.428125 |
| 5 | 1.49 | 8.8 | 192 | 51.2 | 1 | 3300 | 15.6 | 2.044 |

After we have performed a cluster analysis, one of our possible next tasks is to score new observations. In this case, scoring means assigning new observation to existing clusters without rerunning the clustering algorithm. Suppose another public utility, named "Western Montana," is brought to our attention, and it has the following data: coverage = 1; return =5; cost = 150; load = 65; peak = 5; sales = 6000; nuclear = 0; and fuel = 1. We would like to know to which cluster it belongs. So to score this new observation:

1.  Rerun k-means clustering with 5 clusters.

2.  This time, instead of saving the clusters, save the cluster formula. From the red triangle on **K Means NCluster=5**, select **Save Cluster Formula**. Go to the data table, which has 22 rows, right-click in the 23$^{rd}$ row, and select **Add Rows**. For **How many rows to add**, enter **1**.

3.  Click **OK**.

Enter the data for Western Montana in the appropriate cells in the data table. When you type the last datum in fuel and then click in the next cell to enter the company name, a value will appear in the $23^{rd}$ cell of the **Cluster Formula** column: 2. According to the formula created by the *k*-means clustering algorithm, the company Western Montana should be placed in the second cluster. We have scored this new observation.

# K-means versus Hierarchical Clustering

The final solution to the k-means algorithm can depend critically on the initial guess for the k-means. For this reason, it is recommended that k-means be run several times, and these several answers should be compared. Hopefully, a consistent pattern will emerge from the several solutions, and one of them can be chosen as representative of the many solutions. In contrast, for hierarchical clustering, the solution for k clusters depends on the solution for k+1 clusters, and this solution will not change when the algorithm is run again on the same data set.

Usually, it is a good idea to run both algorithms and compare their outputs. Standard bases to choose between the methods are interpretability and usefulness. Does one method produce clusters that are more interesting or easier to interpret? Does the problem at hand lend itself to finding small groups with unusual patterns? Often one method will be preferable on these bases, and the choice is easy.

# References

Few, S. (2006). "Multivariate Analysis Using Parallel Coordinates." Manuscript, www.perceptualedge.com.

Humby, C., T. Hunt, and T. Phillips. (2007). *Scoring Points: How Tesco Continues to Win Customer Loyalty*. 2nd Ed. Philadelphia: Kogan Page.

Johnson, R. A., and D. W. Wichern. (2002). *Applied Multivariate Statistical Analysis*. 5th Ed. Upper Saddle River, NJ: Prentice Hall.

Ying, L., and W. Yuanuan. (2010). "Application of clustering on credit card customer segmentation based on AHP." International Conference on Logistic Systems and Intelligence Management, Vol. 3, 1869–1873.

# Exercises

1. Use hierarchical clustering on the Public Utilities data set. Make sure to use the company name as a label. Use all six methods (*e.g.*, Average, Centroid, Ward, Single, Complete, and Fast Ward), and run each with the data standardized. How many clusters does each algorithm produce?

2. Repeat exercise 1, this time with the data not standardized. How does this affect the results?

3. Use hierarchical clustering on the Freshmen1.jmp data set. How many clusters are there? Use this number to perform a k-means clustering, (Make sure to try several choices k near the one indicated by hierarchical clustering.) Note that k-means will not permit ordinal data. Based on the means of the clusters for the final choice of k, try to name each of the clusters.

4. Use k-means clustering on the churn data set. Try to name the clusters.

# Chapter 8

---

## Decision Trees

## Figure 8.1 A Framework for Multivariate Analysis

| Multivariate Data |
|---|

| DISCOVERY | INTERDEPENDENCE | DEPENDENCE |
|---|---|---|
| Tables | Principal Components/ Factor Analysis | Multiple Regression |
| Graphs | Cluster Analysis | ANOVA |
| | Other Interdependence Techniques | Logistic Regression |
| | | **Decision Trees (8)** |
| | | Neural Networks |
| | | Other Dependence Techniques |

The decision tree is one of the most widely used techniques for describing and organizing multivariate data. As shown in Figure 8.1, a decision tree is one of the dependence techniques in which the dependent variable can be either discrete (the usual case) or continuous. Also, a decision tree is usually considered to be a data mining technique. One of its strengths is its ability to categorize data in ways that other methods cannot. For example, it can uncover nonlinear relationships that might be missed by, say, linear regression. A decision tree is easy to understand and easy to explain, which is always important when an analyst has to communicate results to a non-technical audience. Decision trees do not always produce the best results, but they offer a reasonable compromise between models that perform well and models that can be simply explained. Decision trees are useful not only for modeling, but are also very useful for exploring a data set, especially when you have little idea of how to model the data.

A decision tree is a hierarchical collection of rules that specify how a data set is to be broken up into smaller groups based on a target variable (*i.e.,* a dependent Y variable). If the target variable is categorical, then the decision tree is called a classification tree. If the target variable is continuous, then the decision tree is called a regression tree. In either case, the tree starts with a single root node. Here a decision is made to split the node based on some non-target variable, creating two (or more) new nodes or leaves. Each of these nodes can be similarly split into new nodes, if possible.

Suppose we have historical data on whether bank customers defaulted on small, unsecured personal loans, and we are interested in developing rules to help us decide whether a credit applicant is a good risk or a bad risk. We might build a decision tree as shown in Figure 8.2. Our target variable, risk, is categorical (good or bad), so this is a classification tree. We first break income into two groups: high and low (*e.g.*, income above $75,000 and income below $75,000). Savings is also broken into two groups, high and low. Based on the historical data, persons with high incomes who can easily pay off the loan out of current income have low default rates and are categorized as good risks without regard to their savings. Persons with low incomes cannot pay off the loan out of current income. But they can pay it off out of savings if they have sufficient savings, which explains the rest of the tree. In each "good risk" leaf, the historical data indicate more persons paid back the loan than not. In each "bad risk" leaf, more persons defaulted than not.

**Figure 8.2  Classifying Bank Customers as "Good" or "Bad" Risks for a Loan**

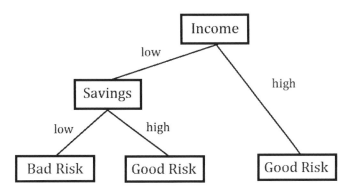

Several questions come to mind. Why only binary splits? Why not split into three or more groups? As Hastie et al. (2009) explain, the reason is that multiway splits use up the data too quickly; besides, multiway splits can always be accomplished by some number of binary splits. You might ask, "What are the cut-off levels for low and high income?" These might be determined by the analyst, who enters the Income variable as categorical, or it might be determined by the computer program if the Income variable is entered as a continuous variable. You might also ask, "Why is income grouped first and then savings? Why not savings first and then income?" The tree-building algorithm determined that splitting first on Income produced "better groups" than splitting first on Savings. Most of the high income customers did not default whereas most of the low income customers defaulted. However, further segregating on the basis of savings revealed that most of the low income customers with high savings did not default.

A common question is, "Can you just put a data set into a tree and expect good results?" The answer is a decided, "No!" Thought must be given to which variables should be included. Theory and business knowledge must guide variable selection. As an example, beware of including a variable that is highly correlated with the target variable. To see the reason for this, suppose the target variable (Y) is whether a prospective customer purchases insurance. If the variable "premium paid" is included in the tree, it will soak up all the variance in the target variable, since only customers who actually do purchase insurance pay premiums. There is also the question of whether an X variable should be transformed and, if so, how. As an example, should Income be entered as continuous, or should it be binned into a discrete variable. If it is to be binned, what should be the cutoffs for the bins? Still more questions remain. If we had many variables, which variables would be used to build the tree? When should we stop splitting nodes?

# An Example of Classification Trees

The questions given above are best answered by means of a simple example. A big problem on college campuses is "freshman retention." For a variety of reasons, many freshmen do not return for their sophomore year. If the causes for departures could be identified, then perhaps remedial programs could be instituted that might enable these students to complete their college education. Open the Freshmen1.jmp data table, which contains 100 observations on several variables that are thought to affect whether a freshman returns for the sophomore year. These variables are described in Table 8.1.

### Table 8.1  The Variables in the freshmen1.jmp Data Set

| Variable | Coding and theoretical reason for including an X variable |
| --- | --- |
| return | =1 if the student returns for sophomore year, =0 otherwise. |
| GPA | Students with a low freshman GPA fail out and do not return. |
| College | The specific college in which the students enroll might affect the decision to return; the engineering college is very demanding and might have a high failure rate. |
| Accommodations | Whether students live in a dorm or on-campus might affect the decision to return. There is not enough dorm space, and some students might hate living off-campus. |
| Part-Time Work Hours | Students who have to work a lot might not enjoy college as much as other students. |
| Attends Office Hours | Students who never attend office hours might be academically weak, need lots of help, and are inclined to fail out; or they might be very strong and never need to attend office hours. |
| High School GPA | Students who were stronger academically in high school might be more inclined to return for sophomore year. |
| Miles from Home | Students who live farther from home might be homesick and wish to transfer to a school closer to home. |

Note that, in the JMP data table, all the variables are correctly labeled as continuous or categorical, and whether they are X or Y variables. Usually, the user has to make these assignments. JMP knows that Return is the target variable (see the small "y" in the blue circle to the right of Return in the columns window), which has 23 zeros and 77 ones. To begin building a decision tree, select **Analyze→Modeling→Partition** and Figure 8.3 will appear.

The data are represented in the graph at the top of the window. Above the horizontal line are 77 data points, and below it (or touching it) are 23 points, which correspond to the number of ones and zeros in the target variable. Points that represent observations are placed in the proper area, above or below the line separating 1 (return) and 0 (not return). As the tree is built, this graph will be subdivided into smaller rectangles, but it can then be hard to interpret. It is much easier to interpret the tree, so we will largely focus on the tree and not on the graph of the data.

**Figure 8.3  Partition Initial Output with Discrete Dependent Variable**

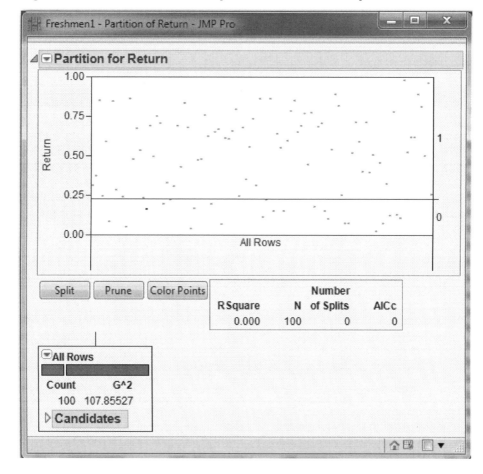

Next notice the box immediately under the graph, which shows information on how well the tree represents the data. Since the tree has yet to be constructed, the familiar RSquare statistic is zero. The number of observations in the data set, N, is 100. No splits have been made, and the AICc is zero. This last statistic is the Akaike Information Criterion where the lowercase C denotes that a correction for finite samples has been applied. The AIC (and AICc) is a model-fitting criterion that trades off bias against variance; a lower AIC (and AICc) is preferred to a higher one.

Next observe the box in the lower left corner, which is the root node of the tree. We can see that there are 100 observations in the data set. The bar indicates how many zeros (red) and ones (blue) are in this node. The "G^2" statistic is displayed; this is the likelihood ratio goodness-of-fit test statistic, and it (like the LogWorth statistic, which is not shown yet) can be used to make splits. What drives the creation of a tree is a criterion function—

*i.e.*, something to be maximized or minimized. In the case of a classification tree, the criterion function by which nodes are split is the LogWorth statistic, which is to be maximized. The Chi-square test of Independence can be applied to the case of multi-way splits and multi-outcome targets. The p-value of the test will indicate the likelihood of a significant relationship between the observed value and the target proportions for each branch. These p-values tend to be very close to zero with large data sets, so the quality of a split is reported by LogWorth = (-1)*ln(chi-squared p-value).

JMP automatically checks all the predictor, or independent, variables, and all the possible splits for them, and chooses the variable and split that maximize the LogWorth statistic. Click the red triangle at the top of the partition window. Under **Display Options**, select **Show Split Prob**. For each level in the data set (zeros and ones), **Rate** and **Prob** numbers have appeared. The **Rate** shows the proportion of each level in the node, while the **Prob** shows the proportion that the model predicts for each level, as shown in Figure 8.4.

### Figure 8.4  Initial Rate, Probabilities, and LogWorths

| All Rows | | |
|---|---|---|
| **Count** | **G^2** | |
| 100 | 107.85527 | |
| **Level** | **Rate** | **Prob** |
| 0 | 0.2300 | 0.2300 |
| 1 | 0.7700 | 0.7700 |

| Candidates | | |
|---|---|---|
| | **Candidate** | |
| **Term** | **G^2** | **LogWorth** |
| GPA | 37.46178905 * | 9.682517958 |
| Miles from Home | 2.22900780 | 0.047507518 |
| College | 8.05280073 | 1.290150751 |
| Accommodations | 0.05564800 | 0.017498340 |
| Part-Time Work Hours | 6.35556865 | 1.132082573 |
| Attends Office Hours | 0.79685672 | 0.209653248 |
| High School GPA | 8.09258865 | 0.929494241 |

Click the gray arrow to expand the **Candidates** box. These are "candidate" variables for the next split. As shown in Figure 8.4, see that GPA has the highest $G^2$ as well as the highest LogWorth. That LogWorth and $G^2$ both are maximized for the same variable is indicated by the asterisk between them. (In the case that $G^2$ and LogWorth are not maximized by the same variable, the "greater than" (>) and "less than" (<) symbols are used to indicate the respective maxima.) In some sense, the LogWorth and $G^2$ statistics can be used to rank the variables in terms of their importance for explaining the target variable, a concept to which we shall return. Clearly, the root node should be split on the GPA variable. Click **Split** to see several changes in the window as shown in Figure 8.5.

**Figure 8.5  Decision Tree after First Split**

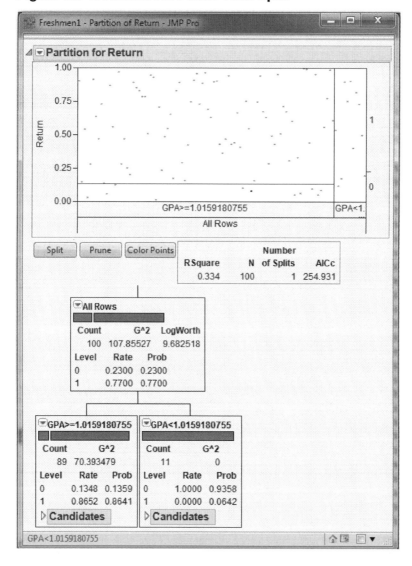

In Figure 8.5, the graph has new rectangles in it, to represent the effect this split has on the data. RSquare has increased from zero to 0.334; the number of splits is 1, and the AICc is 254.9. In the root node, suddenly the LogWorth statistic has appeared. More importantly, the first split divides the data into two nodes depending on whether GPA is above or below 1.0159. The node that contains students whose GPA is less than 1.0159 has 11 students, all of whom are zeros: they did not return for their sophomore year. This node cannot be split anymore because it is pure ($G^2=0$). Splitting the data to create nodes, or subsets, of increasing purity is the goal of tree building. This node provides us with an estimate of the probability that a freshman with a GPA of less than 1.0159 will return for sophomore year: zero—though perhaps the sample size may be a bit small. Perhaps further examination of these students might shed some light on the reasons for their deficient GPAs, and remedial help (extra tutoring, better advising, and specialized study halls) could prevent such students from failing out in the future. In considering the event that a freshman with a GPA above 1.0159 returns for sophomore year, we see that our best estimate of this probability is 0.86 or so. (The simple ratio of returning students to total students in this node is 0.8652, while the model's estimate of the probability is 0.8641.)

The other node, for freshmen with a GPA above 1.0159, contains 89 students. From the blue portion of the bar, we see that most of them are ones (who returned for their sophomore year). Expanding the **Candidates** box in the GPA above 1.0159 node, we see that the largest LogWorth is College, as shown in Figure 8.6.

## Figure 8.6 Candidate Variables for Splitting an Impure Node

GPA>=1.0159180755

| Count | G^2 |
|---|---|
| 89 | 70.393479 |

| Level | Rate | Prob | Count |
|---|---|---|---|
| 0 | 0.1348 | 0.1359 | 12 |
| 1 | 0.8652 | 0.8641 | 77 |

**Candidates**

| Term | Candidate G^2 | | LogWorth |
|---|---|---|---|
| GPA | 18.61968454 | | 3.771459948 |
| Miles from Home | 5.09984838 | | 0.422254132 |
| College | 23.65677128 | * | 4.593602824 |
| Accommodations | 4.79426880 | | 1.164259923 |
| Part-Time Work Hours | 3.55608173 | | 0.519047040 |
| Attends Office Hours | 1.99835034 | | 0.501935536 |
| High School GPA | 6.45939980 | | 0.641257835 |

So the next split will be on the College variable. Click **Split** and expand the **Candidates** boxes in both of the newly created nodes. As can be seen in Figure 8.7, the $R^2$ has increased substantially to 0.553, and the AICc has decreased substantially as well, to 216.145. The parent node has been split into two child nodes, one with 62 students and one with 27 students. The bar for the former indicates one or two zeros—that the node is not pure can be seen by observing that $G^2$= 10.238. (If the node was pure, then $G^2$ would be zero.) The bar for the latter indicates a substantial number of zeros, though still less than half.

### Figure 8.7  Splitting a Node (n=89)

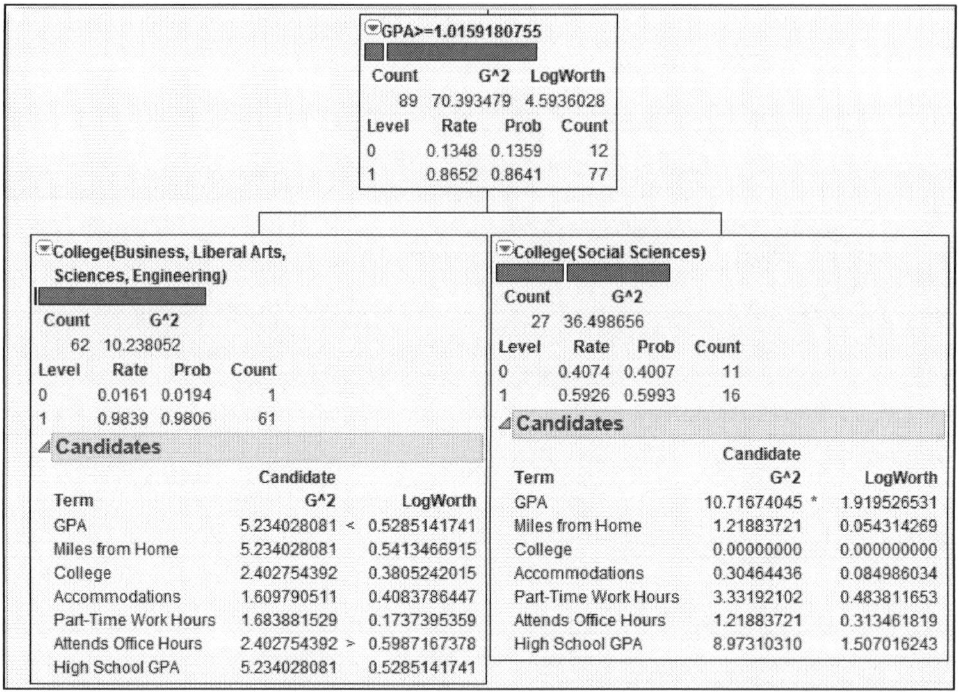

What we see from these splits is that almost all the students who do not return for sophomore year and have not failed out are all in the School of Social Sciences. The schools of Business, Liberal Arts, Sciences, and Engineering do not have a problem retaining students who have not failed out; only one of 62 did not return. What is it about the School of Social Sciences that it leads all the other schools in freshmen not returning?

Before clicking **Split** again, let us try to figure out what will happen next. Check the candidate variables for both nodes. For the node with 62 observations, the largest LogWorth is 0.5987 for the variable **Attends Office Hours**. For the node with 27 observations, the largest LogWorth is 1.919 for the variable **GPA**. We can expect that the node with 27 observations will split next, and it will split on the GPA variable. Look again at the node with 62 observations. We can tell by the Rate for zero that there is only one non-returning student in this group. If there were any doubt, clicking the red triangle in the partitions table, selecting **Display Options**, and then selecting **Show Split Count** shows the count next to the probability. There is really no point in developing further branches from this node, because there is no point in trying to model a 1-out-of-62 occurrence. Click **Split** to see a tree that contains the nodes shown in Figure 8.8.

## Figure 8.8  Splitting a Node (n=27)

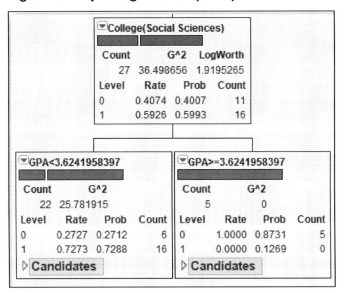

We now have a leaf of non-returning students whose GPA exceeds 3.624. We can conjecture that these students, having achieved a high GPA in their freshman year, are transferring out to better schools. Perhaps they find the school too easy and seek more demanding courses; perhaps the creation of "honors" courses might retain these students. Check the candidate variables for the newly created node for students whose GPA is less than 3.624. The highest LogWorth is 2.4039 for the variable GPA. When we click **Split** again, what will happen? Will the node with 62 observations split, or will the split be on the node for students with a GPA lower than 3.624? Which has the higher LogWorth for a candidate variable? It's the node for students with a GPA lower than 3.624, so that's where the next split will occur. The node with 62 observations will not split until all nodes in the branch for GPA < 3.624 have a LogWorth less than 0.5987. Click **Split**.

**Figure 8.9  Splitting a Node (n=22)**

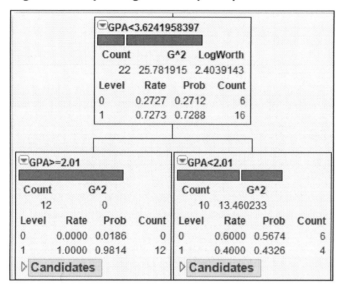

The node for GPA < 3.624 has been split into a leaf for GPA>2.01. These students all return. And a node for GPA < 2.01 is split about 50-50. Check the candidate variables for this node; the highest LogWorth is 0.279 for the variable **Part Time Work Hours**. Since 0.279 is less than 0.5987, the next split will be on the node with 62 observations. But since we're only pursuing a couple of zeros in this branch, there is not much to be gained by analyzing this newest split.

Continue clicking **Split** until the GPA < 2.01 node is split into two nodes with 5 observations each. (This is really far too few observations to make a reliable inference. But let's pretend anyway—30 is a much better minimum number of observations for a node, as recommend by Tuffery, 2011.) The clear difference is that students who work more than 30 hours a week tend to not return with higher probability than students who work fewer than 30 hours a week. More financial aid to these students might decrease their need to work so much, resulting in a higher grade point and a return for their sophomore year. A simple classification tree analysis of this data set has produced a tree with 7 splits and an RSquare of 0.796, and has revealed some actionable insights into improving the retention of freshmen. Note that the algorithm, on its own, found the critical GPA cutoffs of 1.0159 (student has a "D" average) and 2.01 (student has a "C" average).

Suppose that we had not examined each leaf as it was created. Suppose that we built the tree quickly, looking only at $R^2$ or AICc to guide our efforts, and wound up with a tree with several splits, some which we suspect will not be useful. What would we do? We would look for leaves that are predominantly non-returners. This is easily achieved by examining the bars in each node and looking for bars that are predominantly red. Look at the node with 62 observations; this node has but a single non-returning student in it. All the branches and nodes that extend from this node are superfluous, so let us "prune" them from the tree. By pruning the tree, we will end up with a smaller, simpler set of rules that still predicts almost as well as the larger, unpruned tree. Click the red triangle for this node, and click **Prune Below**.

Observe that the RSquare has dropped to 0.764, and the number of splits has dropped to 5. You can see how much simpler the tree has become, without losing any substantive nodes or much of its ability to predict Y. After you grow a tree, it is very often useful to go back and prune nonessential branches. In pruning, we must balance two dangers. A tree that is too large can overfit the data, while a tree that is too small might overlook important structures in the data. There do exist automated methods for growing and pruning trees, but they are beyond the scope of this book. A nontechnical discussion can be found in Linoff and Berry (2011).

Restore the tree to its full size by undoing the pruning. Click the red triangle in the node that was pruned, and click **Split Here**. Next click the just-created Attends Office Hours(Regularly) node and click **Split Here**. The RSquare is again 0.796, and the number of splits is again 7.

Let's conclude this analysis by making use of two more important concepts. First, observe that the seventh and final split only increased the $R^2$ from 0.782 to 0.796. Let us remove the nodes created by the last split. In keeping with the arborist's terminology, click **Prune** to undo the last split, so that the number of splits is 6 and the RSquare is 0.782. Very often we grow a tree larger than we need, and then prune it back to the desired size. After having grown the proper-size classification tree, it is often useful to consider summaries of the tree's predictive power.

The first of these summaries of the ability of a tree to predict the value of Y is to compute probabilities for each observation, as was done for Logistic Regression. Click the red triangle for **Partition for Return** and select **Save Columns→Save Prediction Formula**. In the data table will appear Prob(Return==0), Prob(Return==1) and Most Likely Return. As you look down the rows, there will be much duplication in the probabilities for observations, because all the observations in a particular node are assigned the same probabilities.

Next are two important graphical summaries, the ROC (Receiver Operating Characteristic) curve and the Lift curve. To see these two curves, click the red triangle in the Partition window, click **ROC Curve**, and repeat by clicking **Lift Curve**. See that these curves are now displayed in the output, as shown in Figure 8.10. While they are very useful, for now, we only remark on their existence; they will be explained in great detail in Chapter 9.

**Figure 8.10 ROC and Lift Curves**

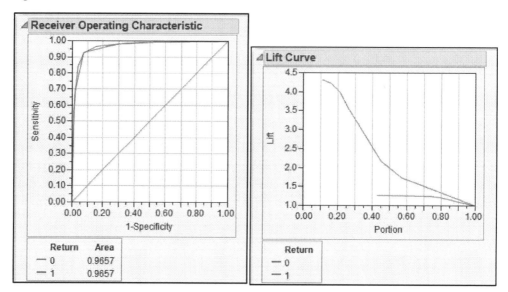

# An Example of a Regression Tree

The type of tree created depends on the target variable. Above we created a classification tree because the Y variable was binary. To use the regression tree approach on the same data set we have been using, we need a continuous target variable.

Suppose we want to explore the components of GPA. In the data table, click the blue x next to GPA and change its role to Y. Click the blue y next to Return and change its role to No Role. As before, select **Analyze**→**Modeling**→**Partition**. See that the root node contains all 100 observations with a mean of 2.216 as shown in Figure 8.11. (If you like, you can verify this by selecting **Analyze**→**Distributions**.) In Figure 8.11, observe the new statistic in the results box, RMSE, which stands for Root Mean Square Error. This can be interpreted as the standard deviation of the target variable in that node. In the present case, GPA is the continuous target variable. Click Candidates to see that the first split will be made on Attends Office Hours. Click **Split**. The two new nodes contain 71 and 29 observations. The former has a mean GPA of 2.00689, and the latter has a mean GPA of 2.728. This is quite a difference in GPA! The regression tree algorithm simply runs a regression of the target variable on a constant for the data in each node, effectively calculating the mean of the target variable in each node.

**Figure 8.11  Partition Initial Output with Continuous Discrete Dependent Variable**

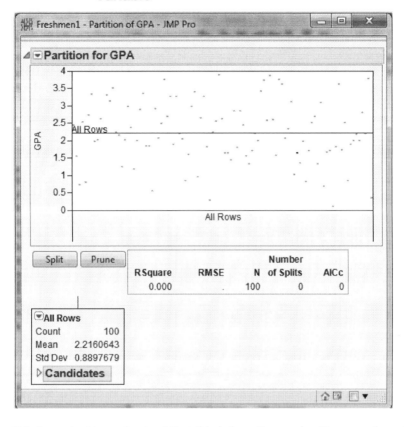

It is important to understand that this is how Regression Trees work, so it is worthwhile to spend some time exploring this idea. We will briefly digress to explore the figure of 2.2160642 that we see in Figure 8.11. Let's create a new variable called Regularly that equals 1 if Attends Office Hours is Regularly and equals zero otherwise.

In the data table, click in the column Attends Office Hours. Select **Cols→Recode**. For Never, change the New Value to 0; for Regularly, change the New Value to 1; and for Sometimes, change the New Value to 0. Select **In Place** and change it to **New Column**. Click **OK** to create a new variable called Attends Office Hours 2. Observe that Attends Office Hours is clearly a nominal variable because it is character data. Attends Office Hours 2 consists of numbers, but should be a nominal variable. Make sure that it is nominal.

To calculate the mean GPA for the Attends Office Hours regularly (Regularly = 1) and Attends Office Hours rarely or never (Regularly = 0), proceed as follows.

Select **Analyze→Distribution**. Under **Y, Columns**, delete all the variables except GPA. Click **Attends Office Hours 2** and then click **By**. Click **OK**. In the Distributions window, you should see that mean GPA is 2.00689 when Attends Office Hours 2 = 0, and that GPA is 2.728 when Attends Office Hours 2 = 1. Regression Trees really do calculate the mean of the target variable for each node. Be sure to delete the variable Attends Office Hours 2 from the data table so that it is not used in subsequent analyses. In the data table, right-click **Attends Office Hours 2** and select **Delete Columns**.

Let's resume with the tree building. Click the red triangle in the first node (with 100 observations) and select **Prune Below**. We are going to split several times, and each time we split we are going to record some information from the RSquare Report Table as shown in Figure 8.12: the RSquare, the RMSE, and the AICc. You can write this down on paper or use JMP. We will use JMP. Right-click in the RSquare Report Table and select **Make into Data Table**. A new data table appears that shows the RSquare, RMSE, N, Number of Split, and AICc. Every time you click **Split**, you produce such a data table. After splitting 14 times, there will be 14 such tables. Copy and paste the information into a single table, as shown in Figure 8.13.

### Figure 8.12 RSquare Report Table before Splitting

| RSquare | RMSE | N | Number of Splits | AICc |
|---------|------|---|------------------|------|
| 0.000 | 0.8853079 | 100 | 0 | 263.547 |

**Figure 8.13 Table Containing Statistics from Several Splits**

| | | RSquare | RMSE | N | Number of Splits | AICc |
|---|---|---|---|---|---|---|
| | 1 | 0.000 | 0.8853078558 | 100 | 0 | 263.54745101 |
| | 2 | 0.137 | 0.8225877196 | 100 | 1 | 250.97767622 |
| | 3 | 0.274 | 0.7543296384 | 100 | 2 | 235.82359521 |
| | 4 | 0.302 | 0.7395153027 | 100 | 3 | 234.07394375 |
| | 5 | 0.356 | 0.7106789006 | 100 | 4 | 228.3840189 |
| | 6 | 0.408 | 0.6813680288 | 100 | 5 | 222.274559 |
| | 7 | 0.429 | 0.6689173837 | 100 | 6 | 220.95118047 |
| | 8 | 0.447 | 0.6583586062 | 100 | 7 | 220.18660627 |
| | 9 | 0.474 | 0.6419799974 | 100 | 8 | 217.61999024 |
| | 10 | 0.490 | 0.631986405 | 100 | 9 | 217.01022741 |
| | 11 | 0.498 | 0.6274553426 | 100 | 10 | 218.157358 |
| | 12 | 0.507 | 0.6215311382 | 100 | 11 | 218.90641127 |
| | 13 | 0.520 | 0.6135277152 | 100 | 12 | 219.02291504 |
| | 14 | 0.527 | 0.6088610506 | 100 | 13 | 220.26895289 |
| | 15 | 0.543 | 0.5983722853 | 100 | 14 | 219.63348991 |

AIC Table - JMP Pro

File Edit Tables Rows Cols DOE Analyze Graph Tools View Window Help

AIC Table

Columns (5/0)
RSquare
RMSE
N
Number of Splits
AICc

Rows
All rows 15
Selected 0
Excluded 0
Hidden 0
Labelled 0

evaluations done

Continue splitting until the number of splits is 14 and the $R^2$ is 0.543. For each split, write down the $R^2$, the RMSE, and the AICc. To use the AICc as a model selection criterion, we would interpret the AICc as a function of the number of splits, with the goal of minimizing the AICc. It sometimes occurs that the AICc simply continues to decrease, perhaps ever so slightly, as the number of splits increases. In such a case, the AICc is not a useful indicator for determining the number of splits. In other cases, the AICc decreases as the number of splits increases, reaches a minimum, and then begins to increase. In such a case, it may be useful to stop splitting when the AICc reaches a minimum. The RMSE and AICc are plotted in Figure 8.14. In the present case, the number of splits that minimizes the AICc is 9, with an AICc of 217.01. It is true that AICc is 219.6 on split 14 and 220.27 on split 13. So the plot of AICc versus splits is not a perfect bowl shape. But it is still true that AICc has a local minimum at 9 splits. All these comments are not to suggest that we cannot continue splitting past 9 splits if we continue to uncover useful information.

**Figure 8.14 Plot of AIC and RMSE versus Number of Splits**

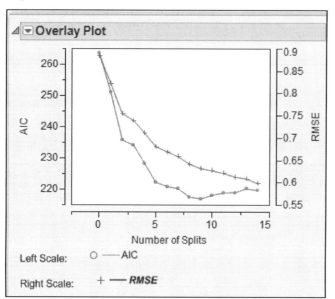

Start over again by pruning everything below the first node. Now create a tree with 9 splits, the number of splits suggested in Figure 8.13. A very practical question is, what can we say about the effect on GPA of the variables in the data set that might be useful to a college administrator? The largest driver of GPA appears to be whether a student attends office hours regularly. This split (and all other splits) effectively divides the node into two groups: high GPA (for this node, students with an average GPA of 2.728) and low GPA (for this node, students with an average GPA of 2.00689). Further, there appears to be no difference between students who attend Never and those who attend Sometimes; we can say this because there is no split between Never and Sometimes students.

For the group of high GPA students (attends office hours regularly with GPA above 2.728), those who live off-campus have substantially lower GPAs than other students. This suggests that the off-campus lifestyle is conducive to many more non-studying activities than living on campus. Among the high GPA students who live in dorms, those who attend school close to home have higher GPAs (3.40 versus 2.74), though the sample size is too small to have much trust in this node. Certainly, if this result was obtained with a larger sample, it might be interesting to investigate the reasons for such a relationship. Perhaps the ability to go home for a non-holiday weekend is conducive to the mental health necessary to sustain a successful studying regimen.

On the low-GPA side of the tree (attends office hours never or sometimes with a GPA of 2.00), we again see that students who attend school closer to home have higher GPAs

(2.83 versus 1.82— *i.e.,* nearly a "B" versus less than a "C"). For those lower-GPA students who live far from home, it really matters whether the student is enrolled in Social Sciences or Liberal Arts on the one hand, or Sciences, Business, or Engineering on the other hand. This might be symptomatic of grade inflation in the former two schools, or it might be a real difference in the quality of the students that enroll in each school. Or it might be some combination of the two. We see also that among low-GPA students in Sciences, Business, and Engineering, more hours of part-time work result in a lower GPA.

Many of these items are interesting, but we wouldn't want to believe any of them—yet. We should run experiments or perform further analyses to document that any effect is real and not merely a chance correlation.

As an illustration of this important principle, consider the credit card company Capital One, which grew from a small division of a regional bank to a behemoth that dominates its industry. Capital One conducts thousands of experiments every year in its attempts to improve the ways that it acquires customers and keeps them. A Capital One analyst might examine some historical data with a tree and find some nodes indicating that blue envelopes with a fixed interest rate offer receive more responses than white envelopes with a variable interest rate offer. Before rolling out such a campaign on a large scale, Capital One would first run a limited experiment to make sure the effect was real, and not just a quirk of the small number of customers in that node of the tree. This could be accomplished by sending out 5,000 offers of each type and comparing the responses within a six-week period.

With a large data set, a tree can grow very large—too large to be viewed on a computer screen. Even if printed out, there are too many leaves to comprehend them all. Let us continue with the Freshmen1.jmp data table, and once again consider the example where the target variable Return is nominal. In the data table, click the blue y next to GPA and make it an X variable. There is no blue letter next to Return, so right-click in the Return column, and select **Preselect Role→Y**. Build a tree with 7 splits as before; the RSquare is 0.796. Under the red triangle in the Partition window, select **Column Contributions**. The chart that appears (see Figure 8.15) tells us which variables are more important in explaining the target variable.

### Figure 8.15 Column Contributions

| Column Contributions | | |
|---|---|---|
| Term | Number of Splits | G^2 |
| GPA | 3 | 60.500211 |
| Miles from Home | 0 | 0.000000 |
| College | 1 | 23.656771 |
| Accommodations | 1 | 1.806975 |
| Part-Time Work Hours | 1 | 1.726092 |
| Attends Office Hours | 1 | 2.402754 |
| High School GPA | 0 | 0.000000 |

If you ever happen to have a large data set and you're not sure where to start looking, build a tree on the data set and look at the **Column Contributions**. (To do this, click the red arrow for the Partition window and then select **Column Contributions**.) In the **Column Contributions** box, right-click anywhere in the SS column (or G^2 column, whichever one appears) and select **Sort by Column**.

Another useful tool for interpreting large data sets is the Leaf Report. To see it, under the red triangle in the Partition window, select **Leaf Report**, which is shown in Figure 8.16. The top set of horizontal bar charts makes it very easy to see which leaves have high and low concentrations of the target variable. The bottom set of horizontal bar charts shows the counts for the various leaves. The splitting rules for each leaf also are shown. With a little practice, you ought to be able to read a Leaf Report almost as easily as you can read a tree.

### Figure 8.16 Leaf Report

| Leaf Report | | |
|---|---|---|
| Leaf Label | Mean | Count |
| Attends Office Hours(Never, Sometimes)&Miles from Home>=143&College(Sciences, Business, Engineering)&Part-Time Work Hours>=10&Accommodations(Other) | 1.05275541 | 7 |
| Attends Office Hours(Never, Sometimes)&Miles from Home>=143&College(Sciences, Business, Engineering)&Part-Time Work Hours>=10&Accommodations(Off-campus, Dorm) | 1.56941891 | 21 |
| Attends Office Hours(Never, Sometimes)&Miles from Home>=143&College(Sciences, Business, Engineering)&Part-Time Work Hours<10 | 2.42094871 | 5 |
| Attends Office Hours(Never, Sometimes)&Miles from Home>=143&College(Social Sciences, Liberal Arts)&Miles from Home<270 | 1.8017779 | 11 |
| Attends Office Hours(Never, Sometimes)&Miles from Home>=143&College(Social Sciences, Liberal Arts)&Miles from Home>=270&Accommodations(Off-campus) | 2.08821519 | 7 |
| Attends Office Hours(Never, Sometimes)&Miles from Home>=143&College(Social Sciences, Liberal Arts)&Miles from Home>=270&Accommodations(Other, Dorm) | 2.69133804 | 7 |
| Attends Office Hours(Never, Sometimes)&Miles from Home<143 | 2.82931398 | 13 |
| Attends Office Hours(Regularly)&Accommodations(Off-campus) | 2.28060709 | 8 |
| Attends Office Hours(Regularly)&Accommodations(Other, Dorm)&Miles from Home>=75 | 2.74050788 | 16 |
| Attends Office Hours(Regularly)&Accommodations(Other, Dorm)&Miles from Home<75 | 3.40482182 | 5 |

The primary drawback of trees is that they are a high-variance procedure; *i.e.*, growing a tree on two similar data sets probably will not produce two similar trees. As an example of a low-variance procedure, if you run the same regression model on two different (but similar) samples, you will likely get similar results. *I.e.*, both regressions will have roughly the same coefficients. In contrast, if you run the same tree on two different (but similar) samples, you will likely get quite different trees. The reason is that in a tree, an error in any one node does not stay in that node but, rather, is propagated down the tree. By this we mean that, if two variables (*e.g.,* Variable A and Variable B) are close contenders for the first split, a small change in the data might affect which of those variables is chosen for the top slit. Splitting on Variable A may well produce a markedly different tree than splitting on Variable B. There are methods such as boosting and bagging to combat this, by growing multiple trees on the same data set and averaging them. But these methods are beyond the scope of this text.

# References

Hastie, Trevor, Robert Tibshirani, and Jerome Friedman. (2009). *The Elements of Statistical Learning: Data Mining, Inference, and Prediction. 2^nd Ed.* New York: Springer. 311.

Linoff, Gordon S., and Michael J. A. Berry. (2011). *Data Mining Techniques: For Marketing, Sales, and Customer Relationship Management. 3^rd ed.* Indianapolis: Wiley Publishing, Inc., Chapter 7.

Tuffery, Stephane. (2011). *Data Mining and Statistics for Decision Making.* Chichester, West Sussex, UK: John Wiley & Sons. 329.

# Exercises

1. Build a classification tree on the churn data set. Remember that you are trying to predict churn, so focus on nodes that have many churners. What useful insights can you make about customers who churn?

2. After building a tree on the churn data set, use the **Column Contributions** to determine which variables may be important. Could these variables be used to improve the Logistic Regression developed in Chapter Five?

3. Build a Regression Tree on the **Masshousing.jmp** data set to predict market value.

# Chapter 9

## Neural Networks

The Neural Networks technique as shown in our multivariate analysis framework in Figure 9.1 is one of the dependence techniques. Neural networks were originally developed to understand biological neural networks and were specifically studied by artificial intelligence researchers to allow computers to develop the ability to learn. In the past 25 years, neural networks have been successfully applied to a wide variety of problems, including predicting the solvency of mortgage applicants, detecting credit card fraud, validating signatures, forecasting stock prices, speech recognition programs, predicting bankruptcies, mammogram screening, determining the probability that a river will flood, and countless others.

**Figure 9.1 A Framework for Multivariate Analysis**

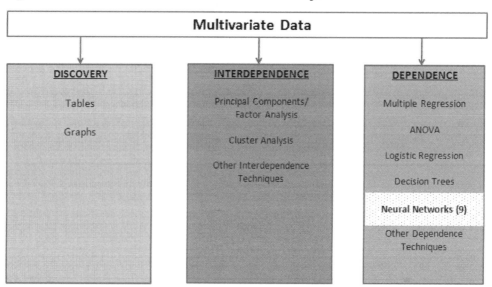

Neural networks are based on a model of how neurons in the brain communicate with each other. In a very simplified representation, a single neuron takes signals/inputs (electrical signals of varying strengths) from an *input* layer of other neurons, weights them appropriately, and combines them to produce signals/outputs (again, electrical signals of varying strengths) as an *output layer*, as shown in Figure 9.2. While this figure shows two outputs in the output layer, most applications of neural networks have only a single output.

## Figure 9.2 A Neuron Accepting Weighted Inputs

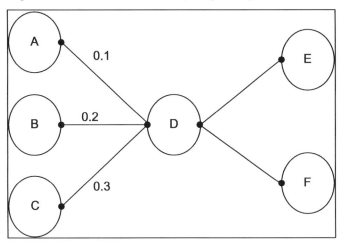

To make things easy, see Figure 9.3 below. Suppose that the strength of each signal emanating from neurons A, B, and C equals one. Now D might accept input from A with a weight of 0.1, from B with a weight of 0.2, and from C with a weight of 0.3. Then the output from D would be 0.1(1) + 0.2(1) + 0.3(1) = 0.6. Similarly, neurons E and F will accept the input of 0.6 with different weights. The linear *activation function* (by which D takes the inputs and combines them to produce an output) looks like a regression with no intercept. If we add what is called a "bias term" of 0.05 to the neuron D as shown in Figure 9.3, then the linear activation function by which D takes the inputs and produces an output is: output =0.05 + 0.1(1) +0.2(1) + 0.3(1) = 0.65.

## Figure 9.3 A Neuron with a Bias Term

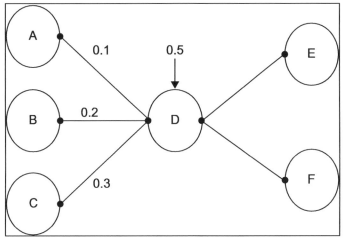

More generally, if Y is the output and the inputs are X1, X2, and X3, then the activation function could be written: Y = 0.05 + 0.1*X1 + 0.2*X2 +0.3*X3. Similarly, neurons E and F would have their own activation functions.

The activation functions used in neural networks are rarely linear as in the above examples, and are usually nonlinear transformations of the linear combination of the inputs. One such nonlinear transformation is the hyperbolic tangent, *tanh*, which would turn the value 0.65 into 0.572 as shown in Figure 9.4. Observe that in the central region of the input, near zero, the relationship between the input and the output is nearly linear; 0.65 is not that far from 0.572. However, farther away from zero, the relationship becomes decidedly nonlinear. Another common activation function is the *Gaussian radial basis function*.

**Figure 9.4  Hyperbolic Tangent Activation Function**

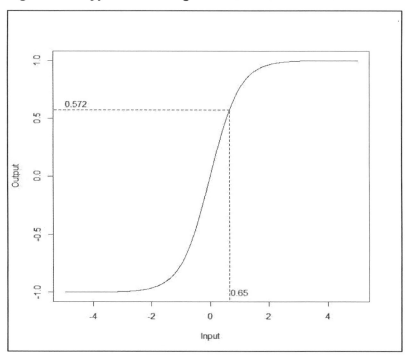

In practice, neural networks are slightly more complicated than those shown in Figures 9.2 and 9.3 and usually look like Figure 9.5 below. Rather than move directly from input to output, to obtain modeling flexibility, the inputs (what we call the X's in a regression problem) are transformed into, say, *features (i.e.,* nodes labeled as Z in the figure below; these are new variables that are combinations of the input variables). Then these variables are used as the inputs that produce the output.

To achieve this goal of flexibility, between the input and output layers is a *hidden layer* that models the features (*i.e.*, creates new variables). As usual, let X represent inputs, Y represent outputs, and let Z represent features. A typical representation of such a neural network is given in Figure 9.5. Let there be k input variables, p features (which means p nodes in the hidden layer), and a single output. Each node of the input layer connects to each node of the hidden layer. Each of these connections has a weight, and each hidden node has a bias term. Each of the hidden nodes has its own activation function that must be chosen by the user.

Similarly, each node of the hidden layer connects to the node in the output layer, and the output node has a bias term. The activation function that produces the output is not chosen by the user: if the output is continuous, it will be a linear combination of the features and if the output is binary, it will be based on a logistic function as discussed in Chapter 5.

## Figure 9.5 A Standard Neural Network Architecture

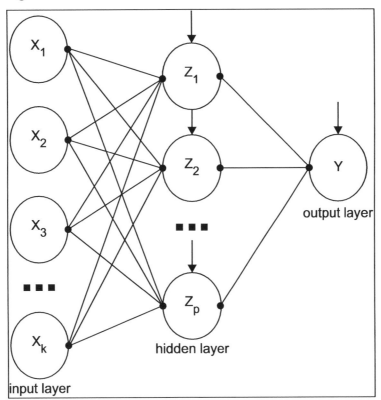

A neural network with k inputs, p hidden nodes, and 1 output has p(k+2)+1 weights to be estimated. So if k= 20 inputs and p=10 hidden nodes, then there are 221 weights to be estimated. How many observations are needed for each weight? It depends on the problem and is, in general, an open question with no definitive answer. But suppose it's 100. Then you need 22,100 observations for such an architecture.

The weights for the connections between the nodes are initially set to random values close to zero and are modified (*i.e.*, trained) on an iterative basis. By default, the criterion for choosing the weights is to minimize the sum of squared errors; this is also called the least squares criterion. The algorithm chooses random numbers close to zero as weights for the nodes, and creates an initial prediction for the output. This initial prediction is compared to the actual output, and the prediction error is calculated. Based on the error, the weights are adjusted, and a new prediction is made that has a smaller sum of squared errors than the previous prediction. The process stops when the sum of squared errors is sufficiently small.

The phrase "sufficiently small" merits elaboration. If left to its own devices, the neural network will make the error smaller and smaller on subsequent iterations, changing the weights on each iteration, until the error cannot be made any smaller. In so doing, the neural network will overfit the model (see Chapter 10 for an extended discussion of this concept) by fitting the model to the random error in the data. Essentially, an overfit model will not generalize well to other data sets. To combat this problem, JMP offers two *validation methods*: holdback and cross-validation.

# Validation Methods

In traditional statistics, and especially in the social sciences and business statistics, one need only run a regression and report an $R^2$ (this is a bit of an over-simplification, but not much). Little or no thought is given to the idea of checking how well the model actually works. In data mining, it is of critical importance that the model "works;" it is almost unheard of to deploy a model without checking whether the model actually works. The primary method for doing this checking is called "validation."

The holdback validation method works in the following way. The data set is randomly divided into two parts, the training sample and the validation (holdback) sample. Both parts have the same underlying model, but each has its own unique random noise. The weights are estimated on the training sample, and these weights are then used on the holdback sample to calculate the error. It is common to use 2/3 of the data for the training and 1/3 for the validation. Initially, as the algorithm iterates, the error on both parts will decline as the neural network learns the model that is common to both parts of the data set. After a sufficient number of iterations (*i.e.*, recalculations of the weights) the neural network will have learned the model, and it will then begin to fit the random noise in the training sample. Since the holdback sample has different random noise, its calculated error will begin to increase.

One way to view this relationship between the error (which we want to minimize) and the number of iterations is displayed in Figure 9.6. After n iterations, the neural network has determined the weights that minimize the error on the holdback sample. Any further iterations will only fit the noise in the training data and not the underlying model (and obviously will not fit the noise in the validation sample). Therefore, the weights based on n iterations that minimize the error on the holdout sample should be used. The curves in Figure 9.6 and the divergence between them as the number of iterations increases is a general method of investigating overfitting.

**Figure 9.6  Typical Error Based on the Training Sample and the Holdback Sample**

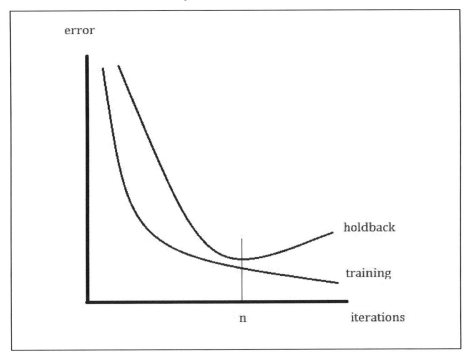

A second way to validate, called *k-fold* cross-validation, works in a different way to determine the number of iterations at which to stop, but the basic idea is the same. Divide the data set into *k* groups, or *folds,* that contain approximately the same number of observations. Consider *k-1* folds to be the training data set, and the *k*th fold to be the validation data set. Compute the relevant measure of accuracy on the validation fold.

Repeat this *k* times, each time leaving out a different fold, thus obtaining *k* measures of accuracy. Average the *k* measures of accuracy to obtain an overall estimate of the accuracy. As with holdback validation, to avoid overfitting, compare the training error to the validation error. Cross-validation is most often used when there are not enough data.

If randomly splitting the data set into two parts is not desirable for some reason, there is a third way to divide the data into training and validation samples. The user can manually split the data into two parts by instructing JMP to include specific rows for the training sample and excluding others that will constitute the holdback sample. There are two ways to do this.

First, the user can select some observations in the data table, right-click, and set them to **Exclude**. Then, when executing a neural net, under **Validation Method**, the user can choose **Excluded Rows Holdback**. Second, the user creates a new variable that indicates by zeros and ones whether a particular observation should be for training (zero) or validation (one). In the Neural dialog box when variables are cast as **Y** or **X**, the user will also set a variable as **Validation**. It will automatically be used for the validation method.

# Hidden Layer Structure

Figure 9.5 shows a neural network architecture with one hidden layer that has *p* nodes. The nodes do not all have to have the same activation function. They can have any combination of the three types of activation functions, tanh, linear, or Gaussian. For example, six nodes in the hidden layer could have all tanh, or could have two each of the three types of the activation functions. Nor is the network limited to one hidden layer: JMP allows up to two hidden layers.

Concerning the architecture of a neural network, the two fundamental questions are: how many nodes for the input layer and how many nodes for the output layer? For the former, the number of variables is the answer. This number cannot be large for two reasons. First, a large number of variables greatly increases the possibility of local optima and correspondingly decreases the probability of finding the global optimum. Second, as the number of variables increases, the amount of time it takes to solve the problem increases even more. A tree or a logistic regression can easily accommodate hundreds of variables. A neural network might not be able to tolerate tens of variables, depending on the sample size and also depending on how long you can wait for JMP to return a solution. The number of nodes in the output layer depends on the problem. To predict a single continuous variable or a binary variable, use one output node. For a categorical variable with five levels, use five output nodes, one for each level.

Suppose there is only a single hidden layer. The guiding principle for determining the number of nodes is that there should be enough nodes to model the target variable, but not so many as to overfit. There are many rules of thumb to follow. Some of which are to set the number of nodes:

- to be between the number of inputs and the number of outputs
- to equal the number of inputs plus the number of outputs times 2/3
- to be no more than twice the number of inputs
- to be approximately ln(T) where T is the sample size
- to be between 5% and 10% of the number of variables

The above list is a set of rules of thumb; notice that the first and second are contradictory. We do not give citations for these rules of thumb because they are all, in some sense, misleading. The fact is that there is no widely accepted procedure for determining the number of nodes in a hidden layer, and there is no formula that will give a single number to answer this question. (Of course, good advice is to find an article that describes successful modeling of the type in question, and use that article as a starting point.)

The necessary number of hidden nodes is a function of, among other things, the sample size, the complexity of the function to be modeled (and this function is usually unknown!), and the amount of noise in the system (which we cannot know unless we know the function!). With too few hidden nodes, the network cannot learn the underlying model and therefore cannot make good predictions on new data; but with too many nodes, the network memorizes the random noise (overfits) and cannot make good predictions on new data. Yoon et al. (1994) indicate that performance improves with each additional hidden node up to a point, after which performance deteriorates.

Therefore, a useful strategy is to begin with some minimum number of hidden nodes and increase the number of hidden nodes, keeping an eye on prediction error via the holdback sample or cross-validation, adding nodes as long as the prediction error continues to decline, and stopping when the prediction error begins to increase. If the training error is low and the validation error (either holdback or k-fold cross-validation) is high, then the model has been overfit, and the number of hidden nodes should be decreased. If both the training and validation errors are high, then more hidden nodes should be added. If the algorithm will not converge, then increase the number of hidden nodes. The bottom line is that extensive experimentation is necessary to determine the appropriate number of hidden nodes. It is possible to find successful neural network models built with as few as three nodes (predicting river flow) and as many as hundreds of hidden nodes (speech and handwriting recognition). The vast majority of analytics applications of neural networks seen by the authors have fewer than thirty hidden nodes.

The number of hidden layers is usually one. The reason to use a second layer is because it greatly reduces the number of hidden nodes that are necessary for successful modeling

(Stathakis, 2009). There is little need to resort to using the second hidden layer until the number of nodes in the first hidden layer has become untenable—*i.e.*, the computer fails to find a solution or takes too long to find a solution. However, using two hidden layers makes it easier to get trapped at a local optimum, which makes it harder to find the global optimum.

*Boosting* is an option that can be used to enhance the predictive ability of the neural network. Boosting is one of the great statistical discoveries of the 20th century. A simplified discussion follows. Consider the case of a binary classification problem (*e.g.*, zero or one). Suppose we have a classification algorithm that is a little better than flipping a coin; *e.g.*, it is correct 55% of the time. This is called a *weak classifier*. Boosting can turn a weak classifier into a strong classifier.

The method of boosting is to run the algorithm once where all the observations have equal weight. In the next step, give more weight to the incorrectly classified observations and less weight to the correctly classified observations, and run the algorithm again. Repeat this re-weighting process until the algorithm has been run T times, where T is the "Number of Models" that the user specifies. Each observation then has been classified T times. If an observation has been classified more times as a zero than a one, then zero is its final classification; if it has been classified more times as a one, then that is its final classification. The final classification model, which uses all T of the weighted classification models, is usually more accurate than the initial classification, sometimes much more accurate. It is not unusual to see the error rate (the proportion of observations that are misclassified) drop from 20% on the initial run of the algorithm to below 5% for the final classification.

Boosting methods have also been developed for predicting continuous target variables. Gradient boosting is the specific form of boosting used for neural networks in JMP, and a further description can be found in the JMP help files. In the example in the manual, a 1-layer/2-node model is run when T=6, and the "final model" has 1 layer and 2x6=12 nodes (individual models are retained and combined at the end), which corresponds to the general method described above in the sense that the T re-weighted models are combined to form the final model.

The method of gradient boosting requires a *learning rate*, which is a number greater than zero and less than or equal to one. The learning rate describes how quickly the algorithm learns: the higher the learning rate, the faster the method converges, but a higher learning rate also increases the probability of overfitting. When the Number of Models (T) is high, then the learning rate should be low, and vice versa.

# Fitting Options

One approach to improving the fit of the model is to transform all the continuous variables to near normality. This can be especially useful when the data contain outliers or are heavily skewed and is recommended in most cases. JMP offers two standard transformation methods, the Johnson Su and Johnson Sb methods. JMP automatically selects the preferred method.

In addition to the default least squares criterion, which minimizes the sum of squared errors, another criterion can be used to choose the weights for the model. The *robust fit* method minimizes the absolute value of the errors rather than the squared errors. This can be useful when outliers are present, since the estimated weights are much more sensitive to squared errors than absolute errors in the presence of outliers.

The *penalty method* combats the tendency of neural networks to overfit the data by imposing a penalty on the estimated weights, or coefficients. Some suggestions for choosing penalties follow. For example, the *squared* method penalizes the square of the coefficients. This is a good choice if you think that most of your inputs contribute to predicting the output; this is the default. The *absolute* method penalizes the absolute value of the coefficients, and this can be useful if you think that only some of your inputs contribute to predicting the output. The *weight decay* method can be useful if you think that only some of your inputs contribute to predicting the output. The *no penalty* option is much faster than the penalty methods, but it usually does not perform as well as the penalty methods because it tends to overfit.

As mentioned previously, to begin the iterative process of estimating the weights for the model, the initial weights are random numbers close to zero. Hence, the final set of weights is a function of the initial, random weights. It is this nature of neural networks to be prone to having multiple local optima that makes it easy for the minimization algorithm to find a local minimum for the sum of squared errors, rather than the global minimum. One set of initial weights can cause the algorithm to converge to one local optimum, while another set of initial weights might lead to another local optimum. Of course, we seek not local optima but the global optimum, and we hope that one set of initial weights might lead to the global optimum.

To guard against finding a local minimum instead of the global minimum, it is customary to restart the model several times using different sets of random initial weights, and then to choose the best of the several solutions. Using the **Number of Tours** option, JMP will do this automatically, so that the user does not need to do the actual comparison by hand. Each "tour" is a start with a new set of random weights. A good number to use is 20 (Sall, Creighton, and Lehman, 2007, p.468).

# Data Preparation

Data with different scales can induce instability in neural networks (Weigend and Gershenfeld, 1994). Even if the network remains stable, having data with unequal scales can greatly increase the time necessary to find a solution. For example, if one variable is measured in thousands and another in units, the algorithm will spend more time adjusting for variation in the former rather than the latter. Two common scales for standardizing the data are:

(1) $\dfrac{x - \overline{x}}{s}$ where $s$ is the sample standard deviation, and

(2) $\dfrac{x - \overline{x}}{\max(x) - \min(x)}$

though other methods exist. Simply for ease, we prefer (1) since it is automated in JMP. (Select **Analyze→Distribution** and after the distribution is plotted, click the red triangle next to the variable name, and select **Save→Standardized**.) Another way is, when creating the neural network model, check the box for **Transform Covariates** under Fitting Options.

In addition to scaling, the data must be scrubbed of outliers. Outliers are especially dangerous for neural networks, because the network will model the outlier rather than the bulk of the data—much more so than, say, with logistic regression. Naturally, outliers should be removed before the data are scaled.

A categorical variable with more than two levels can be converted to dummy variables. But this means adding variables to the model at the expense of making the network harder to train and needing more data.

It is not always the case that categorical variables should be turned into dummy variables. First, especially in the context of neural networks that cannot handle a large number of variables, converting a categorical variable with $k$ categories creates more variables. Doing this conversion with a few categorical variables can easily make the neural network model too difficult for the algorithm to solve. Secondly, conversion to dummy variables can destroy an implicit ordering of the categories that, when maintained, prevents the number of variables from being needlessly multiplied (as happens when converting to dummy variables). Pyle (1999, p. 74) gives an example where a marital status variable with five categories (married, widowed, divorced, single, or never married), instead of being converted to dummy variables, is better modeled as a continuous variable in the [0,1] range as shown in Table 9.1.

**Table 9.1 Converting an Ordered Categorical Variable to [0,1]**

| Never Married | 0 |
|---|---|
| Single | 0.1 |
| Divorced | 0.15 |
| Widowed | 0.65 |
| Married | 1.0 |

This approach is most useful when the levels of the categorical variable embody some implicit order. For example, if the underlying concept is "marriedness," then someone who is divorced has been married more than someone who has never been married. A "single" person may be "never," "divorced," or "widowed," but we don't know which. Such a person is more likely to have experienced marriage than someone who has never been married. See Pyle (1999) for further discussion.

Neural networks are not like a tree or a logistic regression, both of which are useful in their own rights—the former as a description of a data set, the latter as a method of describing the effect of one variable on another. The only purpose of the neural network is for prediction. In the context of predictive analytics, we will generate predictions from a neural network and compare these predictions to those from trees and logistic regression, and then we will use the model that predicts the best. The bases for making these comparisons are the confusion matrix, ROC and lift curves, and various measures of forecast accuracy for continuous variables. All these are explained in the next chapter.

# An Example

The Kuiper.JMP data set comes from Kuiper (2008), and contains the prices of 804 used cars and several variables thought to affect the price. The variable names are self-explanatory, and the interested reader can consult the article (available free and online) for further details. To keep the analysis simple, we will focus on one continuous target variable (Price), two continuous dependent variables (Mileage and Liter) and four binary variables (Doors, Cruise, Sound, and Leather). We ignore the other variables (*e.g.*, Model and Trim). It will probably be easier to run the analyses if variables are defined as "y" or "x."

In the data table, right-click on a variable, select **Preselect Role**, and then select **Y** for Price and **X** for Mileage, Liter, Doors, Cruise, Sound, and Leather.

As a baseline for our neural network modeling efforts, let us first run a linear regression. Of course, before running a regression, we must first examine the data graphically. A scatterplot (select **Graph→Scatterplot Matrix**) shows a clear group of observations near the top of most of the plots. Clicking them indicates that these are observations 151-160. Referring to the data set, we see that these are the Cadillac Hardtop Conv 2D cars. Since these cars appear to be outliers, let us exclude them from the analysis. Select these rows, and then right-click and select **Exclude/Unexclude**. Check the linearity of the relationship (select **Analyze→Fit Y by X**). Click the red arrow on the **Bivariate Fit of Price by Mileage** and select **Fit Line**. Do the same for **Bivariate Fit of Price by Liter**. Linearity seems reasonable, so run the full regression. Select **Analyze→Fit Model** and click **Run**. The RSquare is 0.439881; let's call it 0.44.

With a baseline established, let's turn to the neural network approach. Since we have preselected roles for the relevant variables, they will automatically be assigned as dependent and independent variables. We do not have to select variables each time we run a model. Select **Analyze→Modeling→Neural** and get the object shown in Figure 9.7.

### Figure 9.7 Neural Network Model Launch

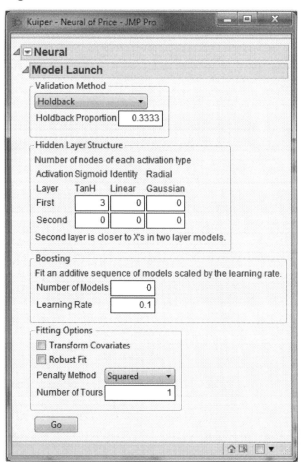

Notice the default options. The validation method is Holdback with a holdback proportion of 1/3. There is a single hidden layer with three nodes, each of which uses the tanh activation function. No boosting is performed, because the **Number of Models** is zero. Neither **Transform Covariates** nor **Robust Fit** is used, though the **Penalty Method** is applied with particular type of method being **Squared**. The model will be run only once, because **Number of Tours** equals one. Click **Go**.

## Figure 9.8 Results of Neural Network Using Default Options

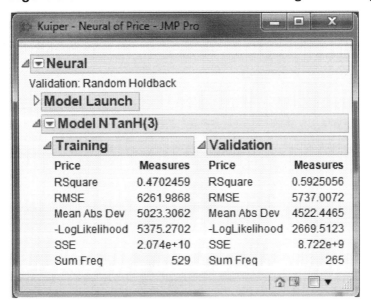

Your results will *not* be exactly like those in Figure 9.8 for two reasons. First, a random number generator is used to determine which 1/3 of the observations were used for the validation sample. Since your training and validation samples differ, so will your numerical results. Even if you had the same training and validation samples, your numerical results would still differ because your initial weights would be different. Remember that the neural network fitting of the weights is an iterative process that must begin somewhere, with some initial weights. These initial weights are random numbers near zero, produced by the JMP random number generator. Since your initial weights are different, your final weights will be different, too.

To get a feel for how much the results can vary, rerun the default neural network nine more times, keeping track of the training and validation RSquare each time. Click the red triangle for Model NTanH(3), and select **Remove Fit**. The Model Launch will reappear; then click **Go**. (You do not have to choose **Remove Fit**, but doing so cuts down on the number of results presented. Alternatively, you can click the triangle for **Model Launch**, and the Model Launch dialog box will reappear.) Our results are presented in Table 2, but remember that your results will differ.

### Table 9.2 Training and Validation RSquare Running the Default Model Ten Times

| Training | Validation | Training | Validation |
|----------|------------|----------|------------|
| 52 | 50 | 59 | 56 |
| 60 | 56 | 55 | 57 |
| 56 | 52 | 62 | 58 |
| 46 | 51 | 61 | 58 |
| 58 | 54 | 56 | 52 |

Notice that the Training RSquare varies substantially, from a low of 46 to a high of 62. These many different solutions are a manifestation of the "multiple local optima" problem that is common to neural networks. Because the model is so highly nonlinear, there often will be several local optima when we really want the unique global optimum. To guard against mistakenly settling for a suboptimal result (*e.g.*, the RSquare of 46 in Table 9.2), it is necessary to run the model several times. Since our goal is to get the best-fitting model on new data, we would naturally choose the model that has the RSquare of 62, as long as it was supported by the Validation RSquare. If the Validation RSquare was much lower than 58, we would take that to be evidence of overfitting and discard the model.

For this reason, JMP has the **Number of Tours** option that will automatically run the model several times and only report the model with the best RSquare. Typically, we expect to see a validation $R^2$ that is near the training $R^2$ if we have successfully modeled the data, or to see a validation $R^2$ that is much below the training $R^2$ if we have overfit the data. Occasionally, we can see a validation $R^2$ that is higher than the training $R^2$, as in the first column of Table 9.2 when the training $R^2$ is 46.

Now that we have some idea of what the default performance is, let's try boosting the default model, setting the **Number of Models** option to 100. Results, shown in Table 9.3, are rather impressive. The average of the 10 training Rsquared values in Table 9.2 is 56.5. The average of the 10 training RSquare values in Table 9.3 is 67.4, an increase of more than 10 points. The downside is that boosting is designed to work with weak models, not strong models. While it can turn a mediocre model into a good model, as we see, it cannot turn a good model into a great model. Usually, one ought to be able to build a model good enough so that boosting won't help. When, however, the best that can be done is a weak model, it's nice to have the boosting method as a backup.

**Table 9.3 Training and Validation RSquare When Boosting the Default Model, Number of Models = 100**

| Training | Validation | Training | Validation |
|----------|------------|----------|------------|
| 71 | 70 | 67 | 65 |
| 67 | 65 | 65 | 65 |
| 66 | 64 | 68 | 66 |
| 66 | 64 | 66 | 62 |
| 69 | 67 | 67 | 65 |

Perhaps other options might improve the performance of the model on the validation data. Table 9.2 (and Table 9.3) shows run-to-run variation from changing initial coefficients for each run. After you click **Go**, the random draws that separate the data into training and validation for any particular run are fixed for the rest of the models that are run in the Neural report window. If this report window is closed and then the model is run again, the training and validation data will be different. To mitigate variation due to changing initial coefficients, we can select the **Number of Tours** option, which tries many different sets of initial coefficients and chooses the run with the best results. With respect to the difference between training and validation runs, we see that sometimes the $R^2$ is approximately the same, and other times one or the other is higher.

In what follows and, indeed, in what we have done so far, we must stress that no firm conclusions can be drawn concerning the superiority of one set of options when the number of runs equals only 5 or 10. There is no reason to rerun the model 100 times in the hope of obtaining definitive conclusions, because the conclusions would apply only to the data set being used. Hence, our conclusions are only tentative and conditional on the data being used.

Next let us compare transforming the variables as well as the method of validation. For this set of experiments, we will set the Number of Tours as 30. Results are presented in Table 9.4, where we use one hidden node with three tanh functions, and compare 10-fold cross-validation with 1/3 holdback sample on the car data, using the transform covariates option with **Number of Tours** =30.

### Table 9.4 R² for Five Runs of Neural Networks

| | k-fold | | holdback | |
|---|---|---|---|---|
| | train | Validate | train | validate |
| **untransformed** | 73 | 62 | 67 | 68 |
| | 64 | 59 | 60 | 57 |
| | 75 | 76 | 73 | 76 |
| | 68 | 52 | 54 | 56 |
| | 74 | 66 | 56 | 55 |
| **transformed** | 71 | 81 | 71 | 62 |
| | 75 | 67 | 71 | 73 |
| | 74 | 88 | 68 | 71 |
| | 73 | 75 | 70 | 73 |
| | 74 | 71 | 68 | 62 |

We shall tentatively conclude that transforming produces better results than not transforming, as evidenced by the higher training $R^2$. (Take care to remember that five runs are far too few to draw such a conclusion.) Look at the untransformed results. If we only ran the model once and happened to get 74/66 (74 for training, 66 for validation) or 73/62, we might think that the architecture we chose overfit the data. Had we happened to get 75/76, we might think we had not overfit the data. This type of random variation in results must be guarded against, which means that the model must be run with many different architectures and many different options. And even then the same architecture with the same options should be run several times to guard against aberrant results.

In the best scenario when we finally have settled on a model and run it several times, all the training $R^2$ would be about the same, all the validation $R^2$ would be about the same, and the validation $R^2$ would be about the same as the training $R^2$ or perhaps a little lower. As for the present case in Table 9.4, we are far from the ideal situation. Between the training and validation data, sometimes the $R^2$ is approximately the same, sometimes it is higher for training, and sometimes it is higher for validation.

We have remarked on situations in which the validation $R^2$ is noticeably higher than the training $R^2$; two such cases occur in Table 9.4. In particular, look at the transformed/k-fold case for 74/88, which has 714 observations in the training data and 80 in the validation data. This situation is usually the result of some observations with large residuals (one hesitates to call them "outliers") that are in the training data not making it into the validation data. See the residuals plots for this case in Figures 9.9a (training data) and 9.9b (validation data). To make such plots, click the red triangle on **Model** and then select **Plot Residual by Predicted**.

**Figure 9.9a Residual Plot for Training Data When R² = 74%**

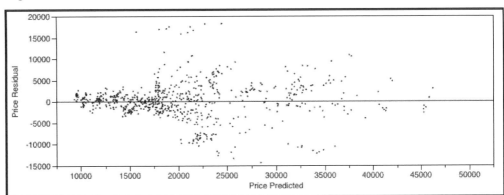

**Figure 9.9b Residual Plot for Validation Data When R² = 88%**

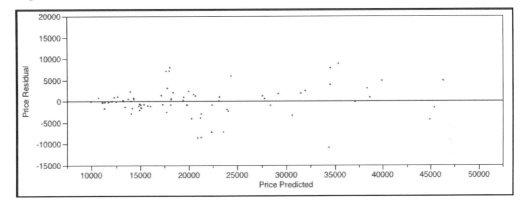

A big difference between these residual plots lies in the 15,000-25,000 range on the x-axis, and in the 15,000-20,000 range on the y-axis. The training data has residuals in this region that the validation data do not, and these are large residuals. Other such differences can be observed (*e.g.*, between 20,000 and 35,000 on the x-axis, below the zero line). Hence, the big difference that we see in the $R^2$ between the two figures.

To be cautious, we should acknowledge that the **Transform Covariates** option might not satisfactorily handle outliers (we may have to remove them manually). Using a robust method of estimation might improve matters by making the estimates depend less on extreme values. For the validation methods, k-fold seems to have a higher $R^2$; this might be due to a deficient sample size, or it might be due to a faulty model.

In Table 9.5 we vary the penalty. We set the **Number of Tours** option to 30, and select **Transform Covariates**, and set **Holdback** to 1/3, one hidden layer with three hidden nodes using the tanh activation function. Clearly, the Squared penalty produces the best results; so, moving forward, we will not use any other penalty for this data set.

### Table 9.5 $R^2$ for Different Penalty Functions

| No Penalty | | Squared | |
|---|---|---|---|
| train | validate | Train | validate |
| 58 | 60 | 73 | 71 |
| 61 | 62 | 74 | 71 |
| 58 | 61 | 75 | 69 |
| 58 | 57 | 73 | 71 |
| 57 | 60 | 75 | 72 |
| **Absolute** | | **Weight Decay** | |
| Train | validate | Train | Validate |
| 58 | 62 | 56 | 60 |
| 53 | 59 | 58 | 60 |
| 53 | 58 | 60 | 62 |
| 61 | 65 | 60 | 62 |
| 58 | 62 | 58 | 61 |

Next we try changing the architecture as shown in Table 9.6. We use the Squared penalty, keep **Number of Tours** at 30, and select the **Transform Covariates** option. We increase the number of hidden nodes from three to four, five, and then ten, and finally change the number of hidden layers to two, each with six hidden nodes. All of these use the tanh function. Using 4 nodes represents an improvement over 3 nodes; and using 5 nodes represents an improvement over 4 nodes. Changing the architecture further produces only marginal improvements.

**Table 9.6 R² for Various Architectures**

| 1 layer, 4 nodes | | 1 layer, 5 nodes | |
|---|---|---|---|
| train | validate | train | validate |
| 75 | 74 | 87 | 83 |
| 79 | 78 | 87 | 84 |
| 76 | 75 | 78 | 77 |
| 75 | 75 | 86 | 84 |
| 85 | 81 | 88 | 85 |
| 1 layer, 10 nodes | | 2 layers, 6 nodes each | |
| train | validate | train | validate |
| 89 | 84 | 91 | 88 |
| 86 | 84 | 87 | 86 |
| 89 | 84 | 87 | 85 |
| 89 | 85 | 89 | 84 |
| 88 | 82 | 89 | 85 |

Compared to linear regression (which had an $R^2$ of 0.44, you may recall), a neural network with one layer and five or ten nodes is quite impressive, as is the network with two layers and six nodes each.

For the sake of completeness, let us briefly consider using the neural network to predict a binary dependent variable. Create a binary variable that equals one if the price is above the median and equals zero otherwise. That the median equals 18205 can be found by selecting **Analyze→Distributions**. To create the binary variable, select **Cols→New Column**, type MedPrice for the column name, click **Column Properties**, and select **Formula**. Under **Functions**, click **Conditional** and select **If**. The relevant part of the dialog box should look like Figure 9.9.

**Figure 9.9 Creating a Binary Variable for Median Price**

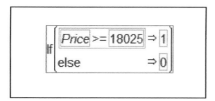

Click **OK**. In the data table, click on the blue y next to Price and change its role to **No Role**. Right-click **MedPrice**, choose **Preselect Role**, and select **Y**. Click the blue triangle next to **MedPrice** and change its type to **Nominal**. Select **Analyze→Modeling→Neural** and click **Go**. To see the confusion matrices (which were discussed briefly in Chapter Five and will be discussed in detail in the next chapter), click the triangles to obtain something like Figure 9.10. (Your results will be numerically different due to the use of the random number generator.)

**Figure 9.10 Default Model for the Binary Dependent Variable, MedPrice**

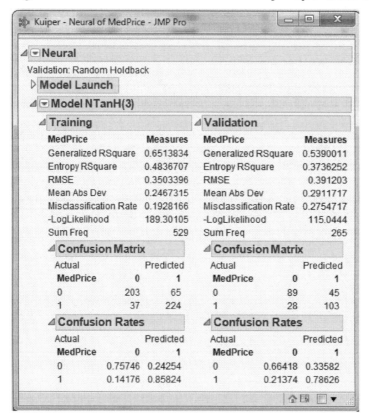

The correct predictions from the model, as shown in the Confusion Matrix, are in the upper left and lower right (*e.g.*, 203 and 224). Looking at the Confusion Rates, we see that the correctly predicted proportions for zeros and ones are 0.75746 and 0.85824, respectively. These rates decline slightly in the validation sample (to 0.66418 and 0.78626), as is typical, because the model tends to overfit the training data.

# Summary

Neural networks are a black box, as far as statistical methods are concerned. The data go in, the prediction comes out, and nobody really knows what goes on inside the network. No hypotheses are tested, there are no p-values to determine whether variables are significant, and there is no way to determine precisely how the model makes its predictions. As such, neural networks, while they may be quite useful to a statistician, are probably not useful when one needs to present a model with results to a management team. The management team is not likely to put much faith in the statistician's presentation: "We don't know what it does or how it does it, but here is what it did." At least trees are intuitive and easily explained, and logistic regression can be couched in terms of relevant variables and hypothesis tests, so these methods are better when one needs to present results to a management team.

Neural networks do have their strong points. They are capable of modeling extremely nonlinear phenomena, require no distributional assumptions, and they can be used for either classification (binary dependent variable) or prediction (continuous dependent variable). Selecting variables for inclusion in a neural network is always difficult, since there is no test for whether a variable makes a contribution to a model. Naturally, consultation with a subject-matter expert can be useful. Another way is to apply a tree to the data, and check the "variable importance" measures from the tree, and use the most important variables from the tree as the variables to be included in the neural network.

# References

Kuiper, Shonda. (2008). "Introduction to Multiple Regression: How Much Is Your Car Worth?" *Journal of Statistics Education*, Vol. 16, No. 3.

Pyle, Dorian. (1999). *Data Preparation for Data Mining*. San Francisco: Morgan Kaufmann.

Sall, John, Lee Creighton, and Ann Lehman. (2007). *JMP Start Statistics: A Guide to Statistics and Data Analysis Using JMP. 4*[th] *Ed.* Cary, NC: SAS Institute Inc.

Stathakis, D. (2009). "How many hidden layers and nodes?" *International Journal of Remote Sensing.* Vol. 30, No. 8, 2133–2147.

Weigend, A. S., and N. A. Gershenfeld (Eds.). (1994). *Time Series Prediction: Forecasting the Future and Understanding the Past.* Reading, MA: Addison-Wesley.

Yoon, Youngohc, Tor Guimaraes, and George Swales. (1994). "Integrating artificial neural networks with rule-based expert systems." *Decision Support Systems*, 11(5), 497–507.

# Exercises

1. Investigate whether the **Robust** option makes a difference. For the Kuiper data that has been used in this chapter (don't forget to drop some observations that are outliers!), run a basic model 20 times (e.g., the type in Table 9.4). Run it 20 more times with the **Robust** option invoked. Characterize the difference between the results, if any. E.g., is there less variability in the $R^2$? Is the difference between training $R^2$ and validation $R^2$ smaller? Now include the outliers, and redo the analysis. Has the effect of the **Robust** option changed?

2. For all the analyses of the Kuiper data in this chapter, ten observations were excluded because they were outliers. Include these observations and rerun the analysis that produced one of the tables in this chapter (e.g., Table 9.4). What is the effect of including these outliers?

3. For the neural net prediction of the binary variable MedPrice, try to find a suitable model by varying the architecture and changing the options.

4. Develop a neural network model for the Churn data.

5. In Chapter 8 we developed trees in two cases: a classification tree to predict whether students return, and a regression tree to predict GPA. Develop a neural network model for each case.

6. As indicated in the text, sometimes rescaling variables can improve the performance of a neural network model. Rescale the variables for an analysis presented in the chapter, or in the exercises, and see whether the results improve.

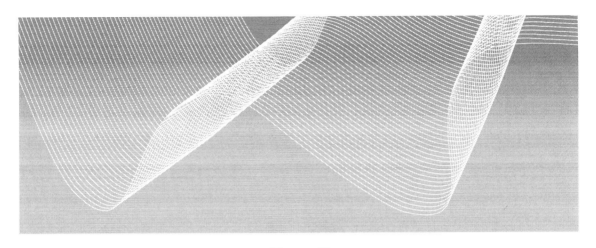

# Chapter 10

---

## Model Comparison

We all know how to compare two linear regression models with the same number of independent variables: look at $R^2$. When the number of independent variables is different between the two regressions, look at adjusted $R^2$. What should you do, though, to compare a linear regression model with a nonlinear regression model, the latter of which really has no directly comparable definition for $R^2$? Suppose that you wish to compare the results of the linear probability model (linear regression applied to a binary variable) to the results of a logistic regression. $R^2$ doesn't work in this case, either.

There is a definite need to compare different types of models so that the better model may be chosen, and that's the topic of this chapter. First, we will examine the case of a continuous dependent variable, which is rather straightforward, if somewhat tedious. Subsequently, we will discuss the binary dependent variable. It permits many different types of comparisons, and its discussion will be quite lengthy.

# Model Comparison with Continuous Dependent Variable

Three common measures used to compare predictions of a continuous variable are the mean square error (MSE), its square root (RMSE), and the mean absolute error (MAE). The last is less sensitive to outliers. All three of these get smaller as the quality of the prediction improves. They are defined as:

MSE:

$$\frac{\sum_{i=1}^{n}(\hat{y}_i - y_i)^2}{n}$$

RMSE:

$$\sqrt{\frac{\sum_{i=1}^{n}(\hat{y}_i - y_i)^2}{n}}$$

MAE:

$$\frac{\sum_{i=1}^{n}|\hat{y}_i - y|}{n}$$

The above performance measures do not consider the level of the variable that is being predicted. For example, an error of 10 units is treated the same regardless of whether the variable has a level of 20 or 2000. To account for the level of the variable, relative measures can be employed, such as:

RELATIVE SQUARED ERROR:

$$\frac{\sum_{i=1}^{n}(\hat{y}_i - y_i)^2}{\sum_{i=1}^{n}(y_i - \bar{y})^2}$$

RELATIVE ABSOLUTE ERROR:

$$\frac{\sum_{i=1}^{n} |\hat{y}_i - y|}{\sum_{i=1}^{n} |y_i - \bar{y}|}$$

The relative measures are particularly useful when comparing variables that have different levels.

Another performance measure is the correlation between the variable and its prediction. Correlation values are constrained to be between -1 and +1 and to increase in absolute value as the quality of the prediction improves. The correlation measure is defined as:

CORRELATION COEFFICIENT:

$$\frac{\sum_{i=1}^{n} (y_i - \bar{y})(\hat{y}_i - \bar{\hat{y}}_i)}{(n-1)s_y s_{\hat{y}}}$$

where $s_y$ and $s_{\hat{y}}$ are the standard deviations of $y$ and $\hat{y}$, respectively, and $\bar{\hat{y}}_i$ is the average of the predicted values.

These performance measures are but a few of many such measures that can be used to distinguish between competing models. Others are the Akaike Information Criterion (AIC), which was discussed in Chapter 8, as well as the similar Bayesian Information Criterion.

In the sequel, we will largely use the absolute rather than the relative measures, because we will be comparing variables that have the same levels.

Which measure should be used can be determined only by a careful study of the problem. Does the data set have outliers? If so, then absolute rather than squared error might be appropriate. If an error of five units when the prediction is 100 is the same as an error of 20 units when the prediction is 400 (*i.e.*, 5% error), then relative measures might be appropriate. Frequently these measures all give the same answer. In that case, it is obvious that one model is not superior to another. On the other hand, there are cases where the measures contradict, and then careful thought is necessary to decide which model is superior.

To explore the uses of these measures, open the file McDonalds48.jmp, which gives monthly returns on McDonalds and the S&P 500 from January 2002 through December 2005. We will run a regression on the first 40 observations and use this regression to make out-of-sample predictions for the last 8 observations. These 8 observations can be called a "hold-out sample."

To exclude the last 8 observations from the regression, select observations 41-48 (click in row 41, hold down the shift key, and click in row 48). Then right-click and select **Exclude/Unexclude**. Each of these rows should have a red circle with a slash through it, similar to [⊘ 41]. Now run the regression:

Select **Analyze→Fit Model**, click **Return on McDonalds**, and then click **Y**. Select **Return on SP500** and click **Add**. Click **Run**. To place the predicted values in the data table, click the red triangle, and select **Save Columns→Predicted Values**.

Notice that JMP has made predictions for observations 41-48, even though these observations were not used to calculate the regression estimates. It is probably easiest to calculate the desired measures using Excel. So either save the datasheet as an Excel file, and then open it in Excel, or just open Excel and copy the variables Return on McDonalds and Predicted Return on McDonalds into columns A and B respectively. In Excel, perform the following steps:

1. Create the residuals in column C, as Return on McDonalds – Predicted Return on McDonalds.

2. Create the squared residuals in column D, by squaring column C.

3. Create the absolute residuals in column E, by taking the absolute value of column C.

4. Calculate the in-sample MSE by summing the first 40 squared residuals, which will be cells 2-41 in column D, and then dividing the sum by 40.

5. Calculate the in-sample MAE by summing the first 40 absolute residuals, which will be cells 2-41 in column E, and then dividing the sum by 40.

6. Calculate the out-of-sample MSE by summing the last 8 squared residuals, cells 42-49 in column D, and then dividing the sum by 8.

7. Calculate the out-of-sample MAE by summing the last 8 absolute residuals, cells 42-49 in column E, and then dividing the sum by 8.

8. Calculate the in-sample correlation between Return on McDonalds and Predicted Return on McDonalds for the first 40 observations using the Excel CORREL( ) function.

9. Calculate the out-of-sample correlation between Return on McDonalds and Predicted Return on McDonalds for the last 8 observations using the Excel CORREL( ) function.

The calculations and results can be found in the file McDonaldsMeasures48.xlsx and are summarized in Table 10.1.

## Table 10.1  Performance Measures for the McDonalds48.jmp File

|               | MSE        | MAE        | Correlation |
|---------------|------------|------------|-------------|
| **In-Sample**     | 0.00338668 | 0.04734918 | 0.68339270  |
| **Out-of-Sample** | 0.00284925 | 0.04293491 | 0.75127994  |

The in-sample and out-of-sample MSE and MAE are quite close, which leads one to think that the model is doing about as well at predicting out-of-sample as it is at predicting in-sample. The correlation confirms this notion. To gain some insight into this phenomenon, look at a graph of Return on McDonalds against Predicted Return on McDonalds.

Select **Graph→Scatterplot Matrix** (or select **Graph→Overlay Plot**), select **Return on McDonalds** and click **Y, Columns**. Then select **Predicted Return on McDonalds** and click **X**. Click **OK**.

In the data table, select observations 41-48, which will make them appear as bold dots on the graph. These can be a bit hard to distinguish. To remedy the problem, while still in the data table, right-click on the selected observations, select **Markers**, and then choose the plus sign (**+**). (See Figure 10.1.) These out-of-sample observations appear to be in agreement with the in-sample observations, which suggests that the relationship between **Y** and **X** that existed during the in-sample period continued through the out-of-sample period. Hence, the in-sample and out-of-sample correlations are approximately the same. If the relationship that existed during the in-sample period had broken down and no longer existed during the out-of-sample period, then the correlations might not be approximately the same, or the out-of-sample points in the scatterplot would not be in agreement with the in-sample points.

**Figure 10.1 Scatterplot with Out-of-Sample Predictions as Plus Signs**

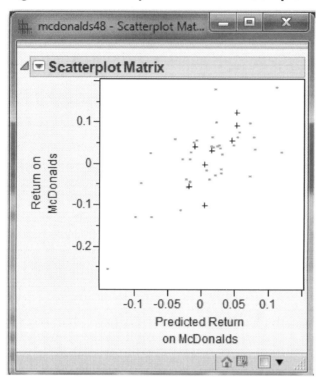

# Model Comparison with Binary Dependent Variable

When actual values are compared against predicted values for a binary variable, a contingency table is used. This table is often called an "error table" or a "confusion matrix." It displays correct versus incorrect classifications, where "1" may be thought of as a "positive/successful" case and "0" may be thought of as a "negative/failure" case.

It is important to understand that the confusion matrix is a function of some threshold score. Imagine making binary predictions from a logistic regression. The logistic regression produces a probability. If the threshold is, say, 0.50, then all observations with a score above 0.50 will be classified as positive, and all observations with a score below 0.50 will be classified as negative. Obviously, if the threshold changes to, say, 0.55, so will the elements of the confusion matrix.

## Table 10.2 An Error Table or Confusion Table

|          | **Predicted 1**        | **Predicted 0**        |
|----------|------------------------|------------------------|
| **Actual 1** | True positive (TP)  | False negative (FN) |
| **Actual 0** | False positive (FP) | True negative (TN)  |

A wide variety of statistics can be calculated from the elements of the confusion matrix, as labeled in Table 10.2. For example, the overall accuracy of the model is measured by the Accuracy:

$$\text{ACCURACY} = \frac{TP + TN}{TP + FP + FN + TN} = \frac{\#\,correctly\,classified}{total\,number\,of\,observations}$$

This is not a particularly useful measure because it gives equal weight to all components. Imagine that you are trying to predict a rare event—say, cell phone churn—when only 1% of customers churn. If you simply predict that all customers do not churn, your accuracy rate will be 99%. (Since you are not predicting any churn, FP=0 and TN=0.) Clearly, better statistics that make better use of the elements of the confusion matrix are needed.

One such measure is the sensitivity, or true positive rate, which is defined as:

$$\text{SENSITIVITY} = \frac{TP}{TP + FN} = \frac{\#\,correctly\,classified\,as\,positive}{\#\,of\,positives}$$

This is also known as *recall*. It answers the question, "If the model predicts a positive event, what is the probability that it really is positive?" Similarly, the true negative rate is also called the *specificity* and is given by:

$$\text{SPECIFICITY} = \frac{TN}{TN + FP} = \frac{\#\,correctly\,classified\,as\,negative}{\#\,of\,negatives}$$

It answers the question, "If the model predicts a negative event, what is the probability that it really is negative?" The false positive rate equals 1-specificity and is given by:

$$\text{FALSE POSITIVE RATE}\,(FPR) = \frac{FP}{TN + FP} = \frac{\#\,incorrectly\,classified\,as\,positive}{\#\,of\,negatives}$$

It answers the question, "If the model predicts a negative event, what is the probability that it is making a mistake?"

When the False Positive Rate (FPR) is plotted on the x-axis, and the True Positive Rate (TPR) is plotted on the y-axis, the resulting graph is called an ROC curve ("Receiver Operating Characteristic Curve"; the name derives from the analysis of radar transmissions in World War II when this graph originated). In order to draw the ROC curve, the classifier has to produce a continuous-valued output that can be used to sort the observations from most likely to least likely. The predicted probabilities from a logistic regression are a good example. In an ROC graph, such as that depicted in Figure 10.2, the vertical axis shows the proportion of ones that are correctly identified, and the horizontal axis shows the proportion of zeros that are misidentified as ones.

**Figure 10.2 An ROC Curve**

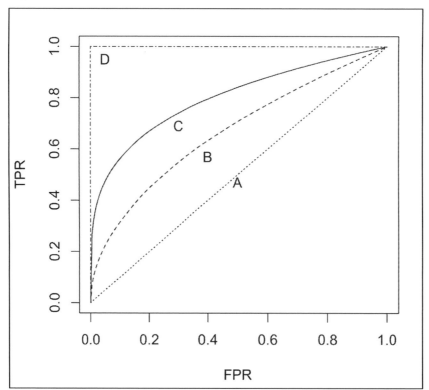

To interpret the ROC curve, first note that the point (0,0) represents a classifier that never issues a positive classification: its FPR is zero, which is good. But it never correctly identifies a positive case, so its TPR is zero, also, which is bad. The point (0,1) represents the perfect classifier: it always correctly identifies positives and never misclassifies a negative as a positive.

In order to understand the curve, two extreme cases need to be identified. First is the random classifier that simply guesses at whether a case is 0 or 1. The ROC for such a classifier is the dotted diagonal line A, from (0, 0) to (1, 1). To see this, suppose that a fair coin is flipped to determine classification. This method will correctly identify half of the positive cases and half of the negative cases, and corresponds to the point (0.5, 0.5). To understand the point (0.8, 0.8), if the coin is biased so that it comes up heads 80% of the time (let "heads" signify "positive"), then it will correctly identify 80% of the positives and incorrectly identify 80% of the negatives. Any point beneath this 45° line is worse than random guessing.

The second extreme case is the perfect classifier, which correctly classifies all positive cases and has no false positives. It is represented by the dot-dash line D, from (0, 0) through (0, 1) to (1, 1). The closer an ROC curve gets to the perfect classifier, the better it is. Therefore, the classifier represented by the solid line C, is better than the classifier represented by the dashed line B. Note that the line C is always above the line B; *i.e.*, the lines do not cross. Remember that each point on an ROC curve corresponds to a particular confusion that, in turn, depends on a specific threshold. This threshold is usually a percentage. *E.g.*, classify the observation as "1" if the probability of its being a "1" is 0.50 or greater. Therefore, any ROC curve represents various confusion matrices generated by a classifier as the threshold is changed. For an example of how to calculate an ROC curve, see Tan, Steinbach, and Kumar (2006, pp. 300–301).

Points in the lower left region of the ROC space identify "conservative" classifiers. They require strong evidence to classify a point as positive. So they have a low false positive rate; necessarily they also have low true positive rates. On the other hand, classifiers in the upper right region can be considered "liberal." They do not require much evidence to classify an event as positive. So they have high true positive rates; necessarily, they also have high false positive rates.

**Figure 10.3 ROC Curves and Line of Optimal Classification**

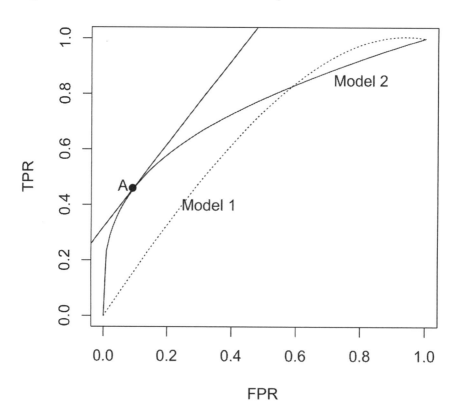

When two ROC curves cross, as they do in Figure 10.3, neither is unambiguously better than the other, but it is possible to identify regions where one classifier is better than the other. Figure 10.3 shows the ROC curve as a dotted line for a classifier produced by Model 1—say, logistic regression—and the ROC curve as a solid line for a classifier produced by Model 2—say, a classification tree. Suppose it is important to keep the FPR low at 0.2. Then, clearly, Model 2 would be preferred because when FPR is 0.2, it has a much higher TPR than Model 1. Conversely, if it was important to have a high TPR— say, 0.9—then Model 1 would be preferable to Model 2 because when TPR =0.9, Model 1 has an FPR of about 0.7 while Model 2 has an FPR of about 0.8.

Additionally, the ROC can be used to determine the point with optimal classification accuracy. Straight lines with equal classification accuracy can be drawn, and these lines will all be from the lower left to the upper right. The line that is tangent to an ROC curve marks the optimal point on that ROC curve. In Figure 10.2, the point marked A for Model 2, with an FPR of about 0.1 and a TPR of about 0.45, is an optimal point. Precise details for calculating the line of optimal classification can be found in Vuk and Curk

(2006, Section 4.1). This point is optimal assuming that the costs of misclassification are equal, that a false positive is just as harmful as a false negative. This assumption is not always true, as shown by misclassifying the issuance of credit cards. A good customer might charge $5000 per year and carry a monthly balance of $200, resulting in a net profit of $100 to the credit card company. A bad customer might run up charges of $1000 before his card is canceled. Clearly, the cost of refusing credit to a good customer is not the same as the cost of granting credit to a bad customer.

A popular method for comparing ROC curves is to calculate the "Area Under the Curve" (AUC). Since both the x-axis and y-axis are from zero to one, and since the perfect classifier passes through the point (0,1), the largest AUC is one. The AUC for the random classifier (the diagonal line) is 0.5. In general, then, an ROC with a higher AUC is preferred to an ROC curve with a lower AUC. The AUC has a probabilistic interpretation. It can be shown that AUC = P(random positive example > random negative example). In words, this is the probability that the classifier will assign a higher score to a randomly chosen positive case than to a randomly chosen negative case.

To illustrate the concepts discussed here, let us examine a pair of examples with real data. We will construct two simple models for predicting churn, and then compare them on the basis of ROC curves. Open the churn data set (from Chapter 5) and fit a logistic regression. Churn is the dependent variable (make sure that it is classified as nominal), where TRUE or 1 indicates that the customer switched carriers. For simplicity, choose D_VMAIL_PLAN, VMail_Message, Day_Mins, and Day_Charge as explanatory variables and leave them as continuous. Click **Run**. Under the red triangle for the Nominal Logistic Fit window, click **ROC Curve**. Since we are interested in identifying churners, when the pop-up box instructs you to **Select which level is the positive**, select **1** and click **OK**.

**Figure 10.4 ROC Curves for Logistic (Left) and Partition (Right)**

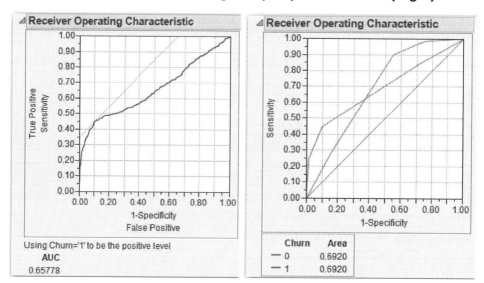

Observe the ROC curve together with the line of optimal classification; the AUC is 0.65778, the left ROC curve in Figure 10.4. The line of optimal classification appears to be tangent to the ROC at about 0.10 for 1-Specificity and about 0.45 for Sensitivity. At the bottom of the window is a tab for the **ROC Table**. Expand it to see various statistics for the entire data set. Imagine a column between Sens-(1-Spec) and True Pos; scroll down until Prob = 0.2284, and you will see an asterisk in the imagined column. This asterisk denotes the row with the highest value of Sensitivity (1-Specificity), which is the point of optimal classification accuracy. Should you happen to have 200,000 rows, right-click in the **ROC Table** and select **Make into Data Table**, which will be easy to manipulate to find the optimal point. Let's try it on the present example.

In the Logistic Table beneath the ROC Curve, click to expand **ROC Table**. Right-click in the table itself and select **Make into Data Table.** In the data table that appears, Column 5 is the imaginary column (JMP creates it for you). Select **Rows→Data Filter**. Select Column 5 and click **Add**. In the Data Filter that appears (see Figure 10.5), select the asterisk by clicking the box with the asterisk. Close the data filter by clicking the red **X** in the upper right corner.

### Figure 10.5 Data Filter

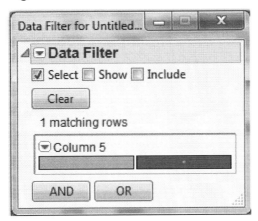

In the data table that you have created, select **Rows→Next Selected** to go to Row 377, which contains the asterisk in Column 5.

We want to show how JMP compares models, so we will go back to the churn data set and use the same variables to build a classification tree. Select **Analyze→Modeling→Partition**. Use the same Y and X variables as for the logistic regression (Churn versus D_VMAIL_PLAN, VMail_Message, Day_Mins, and Day_Charge). Click **Split** five times so that RSquare equals 0.156. Under the red triangle for the Partition window, click **ROC Curve**. This time you are not asked to select which level is positive; you are shown two ROC Curves, one for False and one for True, as shown in the right ROC curve in Figure 10.3. They both have the same AUC because they represent the same information. Observe that one is a reflection of the other. Note that the AUC is 0.6920. The partition method does not produce a line of optimal classification because it does not produce an ROC Table. On the basis of AUC, the classification tree seems to be marginally better than the logistic regression.

# Model Comparison Using the Lift Chart

Suppose you intend to send out a direct mail advertisement to all 100,000 of your customers and, on the basis of experience, you expect 1% of them to respond positively (*e.g.*, to buy the product). Suppose further that each positive response is worth $200 to your company. Direct mail is expensive; it will cost $1 to send out each advertisement. You expect $200,000 in revenue, and you have $100,000 in costs. Hence you expect to make a profit of $100,000 for the direct mail campaign. Wouldn't it be nice if you could send out 40,000 advertisements from which you could expect 850 positive responses?

You would save $60,000 in mailing costs and forego $150x200=$30,000 in revenue for a profit of $170,000-$40,000=$130,000.

The key is to send the advertisement only to those customers most likely to respond positively, and not to send the advertisement to those customers who are not likely to respond positively. A logistic regression, for example, can be used to calculate the probability of a positive response for each customer. These probabilities can be used to rank the customers from most likely to least likely to respond positively. The only remaining question is how many of the most likely customers to target.

A standard lift chart is constructed by breaking the population into deciles, and noting the expected number of positive responses for each decile. Continuing with the direct mail analogy, we might see lift values as shown in Table 10-3:

**Table 10.3 Lift Values**

| decile | customers | responses | response rate | lift |
|--------|-----------|-----------|---------------|------|
| 1 | 10,000 | 280 | 2.80 | 2.80 |
| 2 | 10,000 | 235 | 2.35 | 2.35 |
| 3 | 10,000 | 205 | 2.05 | 2.05 |
| 4 | 10,000 | 130 | 1.30 | 1.30 |
| 5 | 10,000 | 45 | 0.45 | 0.45 |
| 6 | 10,000 | 35 | 0.35 | 0.35 |
| 7 | 10,000 | 25 | 0.25 | 0.25 |
| 8 | 10,000 | 20 | 0.20 | 0.20 |
| 9 | 10,000 | 15 | 0.15 | 0.15 |
| 10 | 10,000 | 10 | 0.10 | 0.10 |
| totals | 100,000 | 1,000 | 1 | |

If mailing was random, we would expect to see 100 positive responses in each decile. (The overall probability of "success" is 1,000/100,000 = 1%, and the expected number of successful mailings in a decile is 1% of 10,000 = 100.) However, since the customers were scored (had probabilities of positive response calculated for each of them), we can expect 280 responses from the first 10,000 customers. Compared to the 100 that would be achieved by random mailing, scoring gives a "lift" of 280/100 = 2.8 for the first decile. Similarly, the second decile has a lift of 2.35.

A lift chart does the same thing, except on a more finely graduated scale. Instead of showing the lift for each decile, it shows the lift for each percentile. Necessarily, the lift for the 100th percentile equals one. Consequently, even a poor model lift is always equal to or greater than one.

To create a lift chart, refer back to the previous section in this chapter where we produced a simple logistic regression and a simple classification tree. This time, instead of selecting **ROC Curve**, select **Lift Curve**. It is difficult to compare graphs when they are not on the same scale, and, further, we cannot see the top of the Lift Curve for the classification tree. See Figure 10.6.

**Figure 10.6  Initial Lift Curves for Logistic (Left) and Classification Tree (Right)**

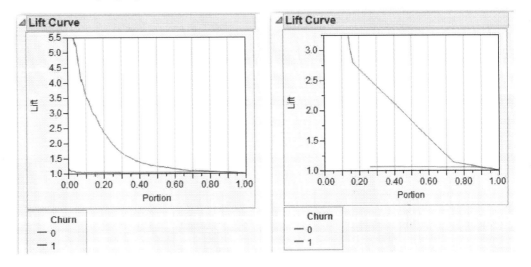

Let's extend the y-axis for both curves to 6.0. Right-click **Lift Curve** for the classification tree and select **Size/Scale→Y Axis**. Near the top of the pop-up box, change the Maximum from 3.8 to 6. Click **OK**. Do the same thing for the **Logistic Lift Curve**. (Alternatively, if you have one axis the way you like it, you can right-click it and select **Edit→Copy Axis Settings**. Then go to the other graph, right-click on the axis, and select **Edit→Paste Axis Settings**.)

Both lift curves, in Figure 10.7, show two curves, one for False and one for True. We are obviously concerned with True, since we are trying to identify churners. Suppose we wanted to launch a campaign to contact customers who are likely to churn, and we want to offer them incentives not to churn. Suppose further that, due to budgetary factors, we could contact only 40% of them. Clearly we would want to use the classification tree, because the lift is so much greater in the range 0 to 0.40.

**Figure 10.7 Lift Curves for Logistic (Left) and Classification Tree (Right)**

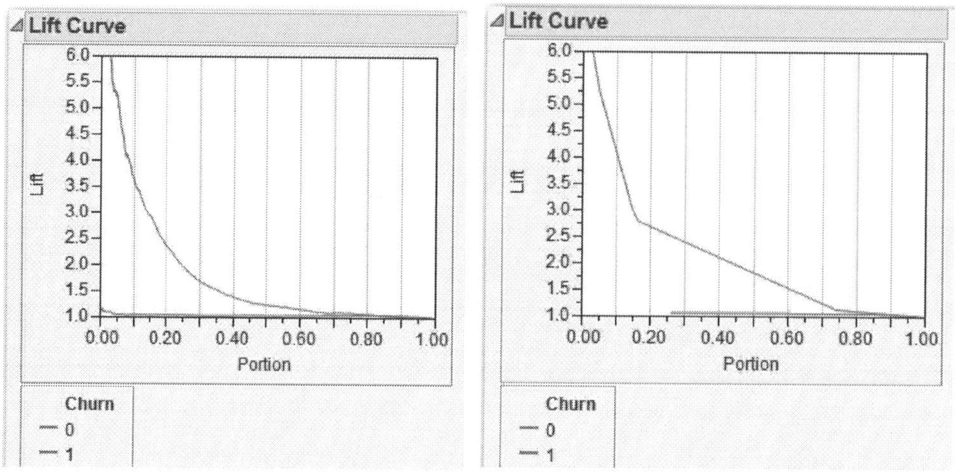

# Train, Validate, and Test

It is common in the social sciences and in some business settings to build a model and use it without checking whether the model actually works. In such situations, the model often overfits the data; *i.e.*, it is unrealistically optimistic because the analyst has fit not just the underlying model but also the random errors. While the underlying model may persist into the future, the random errors will definitely be different in the future. In data mining, when real money is on the line, such an approach is a recipe for disaster.

Therefore, data miners typically divide their data into three sets: a training set, a validation set, and a test set. The training set is used to develop different types of models. For example, an analyst might estimate twenty different logistic models before settling on the one that works the best. Similarly, the analyst might build 30 different trees before finding the best one. In both cases, the model probably will be overly optimistic. Rather than compare the best logistic and best tree based on the training data, the analyst should then compare them on the basis of the validation data set, and choose the model that performs best. Even this will be somewhat overly optimistic. So to get an unbiased assessment of the model's performance, the model that wins on the validation data set should be run on the test data set.

To illustrate these ideas, open McDonalds72.jmp, which contains the monthly returns on McDonald's stock, the monthly return on the S&P 500, and the monthly returns on 30 other stocks, for the period January 2000 through December 2005. We will analyze the first 60 observations, using the last 12 as a holdout sample.

As we did earlier in this chapter, select observations 61-72, right-click, and select **Exclude/Unexclude**. Then select **Fit Y by X** and click the red triangle to regress McDonalds (**Y, response**) on the S&P 500 (**X, factor**). Observe that RSquared is an anemic 0.271809, which, since this is a bivariate regression, implies that the correlation between $y$ and $\hat{y}$ is $\sqrt{0.27189} = 0.52413$. This can easily be confirmed. Select **Analyze→Multivariate Methods→Multivariate**. Then select **Return on McDonalds** and **Return on SP500**, click **Y, Columns**, and click **O**.

Surely some of the other 30 stocks in the data set could help improve the prediction of McDonald's monthly returns. Rather than check each stock manually, we will use stepwise regression to automate the procedure:

Select **Analyze→Fit Model**. Click **Return on McDonalds** and click **Y**. Click **Return on S&P 500** and each of the other thirty stocks. Then click **Add**. Under Personality, click **Stepwise**. Click **Run**. The Fit Stepwise page will open.

Under Stepwise Regression Control, for the Stopping Rule, select **P-value Threshold.** Observe that Prob to enter is 0.25 and Prob to leave is 0.1. For Direction, select **Mixed.** Note that Prob to enter is still 0.25, but Prob to leave is now 0.25. Change Prob to leave back to 0.1.

Next to each variable are the options **Lock** and **Entered**. **Entered** will include a variable, but it might be dropped later. To keep it always, select **Lock** after selecting **Entered**. If you want a variable always omitted, then leave **Entered** blank and check **Lock**. We always want **Return on SP500** in the model, so check **Entered** and then **Lock** for this variable. (See Figure 10.8.) Click **Go**.

### Figure 10.8 Control Panel for Stepwise Regression

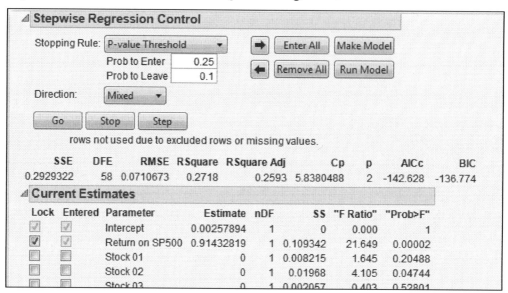

Observe that all the p-values (Prob>F) for the included variables (checked variables) in the stepwise output (not counting the intercept) are less than 0.05 except for Stock 21. So uncheck the box next to Stock 21 and click **Run Model** at the top of the Control Panel. The regression output for the selected model then appears. See Figure 10.9.

**Figure 10.9  Regression Output for Model Chosen by Stepwise Regression**

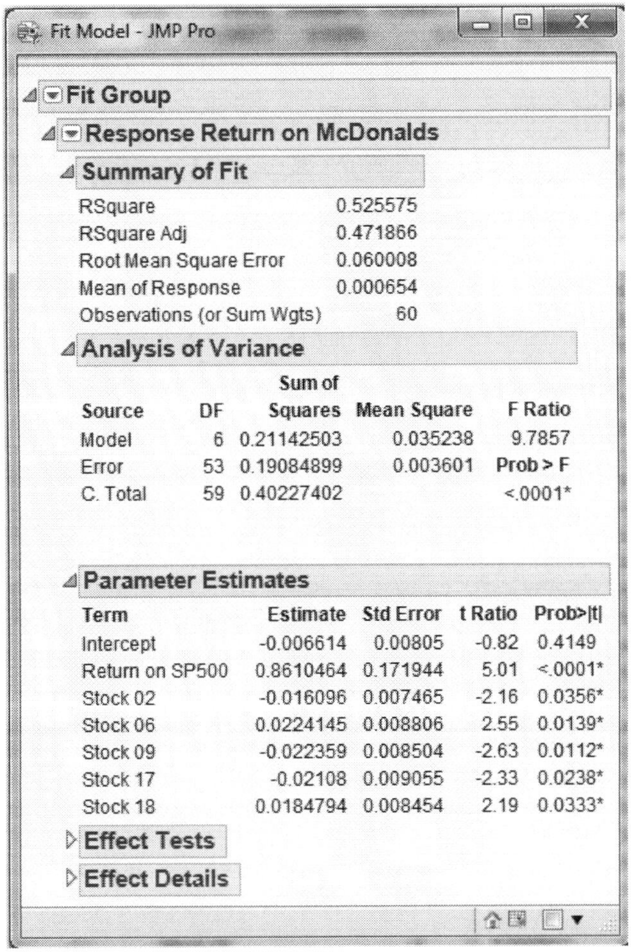

Fit Model - JMP Pro

**Fit Group**

**Response Return on McDonalds**

**Summary of Fit**

| | |
|---|---|
| RSquare | 0.525575 |
| RSquare Adj | 0.471866 |
| Root Mean Square Error | 0.060008 |
| Mean of Response | 0.000654 |
| Observations (or Sum Wgts) | 60 |

**Analysis of Variance**

| Source | DF | Sum of Squares | Mean Square | F Ratio |
|---|---|---|---|---|
| Model | 6 | 0.21142503 | 0.035238 | 9.7857 |
| Error | 53 | 0.19084899 | 0.003601 | Prob > F |
| C. Total | 59 | 0.40227402 | | <.0001* |

**Parameter Estimates**

| Term | Estimate | Std Error | t Ratio | Prob>|t| |
|---|---|---|---|---|
| Intercept | -0.006614 | 0.00805 | -0.82 | 0.4149 |
| Return on SP500 | 0.8610464 | 0.171944 | 5.01 | <.0001* |
| Stock 02 | -0.016096 | 0.007465 | -2.16 | 0.0356* |
| Stock 06 | 0.0224145 | 0.008806 | 2.55 | 0.0139* |
| Stock 09 | -0.022359 | 0.008504 | -2.63 | 0.0112* |
| Stock 17 | -0.02108 | 0.009055 | -2.33 | 0.0238* |
| Stock 18 | 0.0184794 | 0.008454 | 2.19 | 0.0333* |

▷ **Effect Tests**

▷ **Effect Details**

We now have a much higher RSquared of 0.525575 and five stocks (in addition to S&P500) that contribute to explaining the variation in the Return on McDonalds. Stocks 02, 06, 09, 17, and 18 all have p-values of less than 0.05. In all, we have a very respectable regression model; we shouldn't expect to get an RSquared of 0.9 or 0.95 when trying to explain stock returns.

Indeed, this is where many such analyses stop: with a decent $R^2$ and high t-stats on the coefficients. Concerned as we are with prediction, we have to go further and ask, "How well does this model predict?" If we have correctly fitted the model, then we should expect to see an $R^2$ of about 0.53 on the holdout sample; this would correspond to a correlation between predicted and actual of $\sqrt{0.53} \approx 0.73$.

We can compute the MSE, MAE, and Correlation for both in-sample and out-of-sample as shown in Table 10.4. In the Fit Model window, click the red triangle next to **Response Return on McDonalds** and select **Save Columns→Predicted Values**. Then follow the same steps (outlined earlier in the chapter) that were used to create Table 10.1. An Excel spreadsheet for the calculations is McDonaldsMeasures72.xlsx. You will find:

### Table 10.4  Performance Measures for the McDonalds72.jmp File

|  | MSE | MAE | Correlation |
|---|---|---|---|
| **In-Sample** | 0.00318082 | 0.04520731 | 0.72496528 |
| **Out-of-Sample** | 0.00307721 | 0.04416216 | 0.52530075 |

(If you have trouble reproducing Table 10.4, see the steps at the end of the exercises.)

MSE and MAE are commonly used to compare in-sample and out-of-sample data sets, but they can be misleading. In this case, the MSE and MAE for both in-sample and out-of-sample appear to be about the same, but look at the correlations. The in-sample correlation of 0.725 compares with the RSquared of the model. Yet the out-of-sample correlation is the same as the original bivariate regression. What conclusion can we draw from this discrepancy? It is clear that the five additional stocks boost only the in-sample $R^2$ and have absolutely no effect on out-of-sample. How can this be?

The reason is that the additional stocks have absolutely no predictive power for McDonald's monthly returns. The in-sample regression is simply fitting the random noise in the 30 stocks, not the underlying relationship between the stocks and McDonalds. In point of fact, the 30 additional stock returns are not really stock returns, but random numbers that are generated from a random normal distribution with mean zero and unit variance.[1] To see other examples of this phenomenon, consult the article by Leinweber (2007). Let's take another look at this overfitting phenomenon using coin flips.

Imagine that you have forty coins, some of which may be biased. You have ten each of pennies, nickels, dimes, and quarters. You do not know the bias of each coin, but you want to find the coin of each type that most often comes up heads. You flip each coin fifty times and count the number of heads to get the results shown in Table 10.5.

## Table 10.5  The Number of Heads Observed When Each Coin Was Tossed 50 Times

| | Coin (number of heads observed in 50 tosses) | | | | | | | | | |
|---|---|---|---|---|---|---|---|---|---|---|
| | **1** | **2** | **3** | **4** | **5** | **6** | **7** | **8** | **9** | **10** |
| Penny | 21 | 27 | 25 | 28 | 26 | 25 | 19 | <u>32</u> | 26 | 27 |
| Nickel | 22 | 29 | 25 | 17 | <u>31</u> | 22 | 25 | 23 | 29 | 20 |
| Dime | 28 | 23 | 24 | 23 | <u>33</u> | 18 | 22 | 19 | 29 | 28 |
| Quarter | 27 | 17 | 24 | 26 | 22 | 26 | 25 | 22 | 28 | 21 |

Apparently, none of the quarters is biased toward coming up heads. But one penny comes up heads 32 times (64% of the time); one nickel comes up heads 31 times (62%); and one dime comes up heads 33 times (66%). You are now well-equipped to flip coins for money with your friends, having three coins that come up heads much more often than random. As long as your friends are using fair coins, you will make quite a bit of money. Won't you?

Suppose you want to decide which coin is most biased. You flip each of the three coins 50 times and get the results shown in Table 10.6.

## Table 10.6  The Number of Heads in 50 Tosses with the Three Coins We Believe to be Biased

| | |
|---|---|
| **Penny #8** | 26 |
| **Nickel #5** | <u>32</u> |
| **Dime #5** | 28 |

Maybe the penny and dime weren't really biased, but the nickel certainly is. Maybe you'd better use this nickel when you flip coins with your friends. You use the nickel to flip coins with your friends, and, to your great surprise, you don't win any money. You don't lose any money. You break even. What happened to your nickel? You take your special nickel home, flip it 100 times, and it comes up heads 51 times. What happened to your nickel?

In point of fact, each coin was fair, and what you observed in the trials was random fluctuation. When you flip ten separate pennies fifty times each, some pennies are going to come up heads more often. Similarly, when you flip a penny, a nickel, and a dime fifty times each, one of them is going to come up heads more often.

We know this is true for coins, but it's also true for statistical models. If you try 20 different specifications of a logistic regression on the same data set, one of them is going to appear better than the others if only by chance. If you try 20 different specifications of a classification tree on the same data set, one of them is going to appear better than the others. If you try 20 different specifications of other methods like discriminant analysis, neural networks, and nearest neighbors, you have five different types of coins, each of which has been flipped 20 times. If you take a new data set and apply it to each of these five methods, the best method probably will perform the best. But its success rate will be overestimated due to random chance. To get a good estimate of its success rate, you will need a third data set to use on this one model.

Thus, we have the train-validate-test paradigm for model evaluation to guard against overfitting. In data mining, we almost always have enough data to split it into three sets. (This is not true for traditional applied statistics, which frequently has small data sets.) For this reason, we split our data set into three parts: training, validating, and testing. For each statistical method (*e.g*, linear regression and regression trees), we develop our best model on the training data set. Then we compare the best linear regression and the best regression tree on the validation data set. This comparison of models on the validation data set is often called a horserace. Finally, we take the winner (say, linear regression) and apply it to the test data set to get an unbiased estimate of its $R^2$, or other measure of accuracy.

# References

Foster, Dean P., and Robert A. Stine. (2006). "Honest Confidence Intervals for the Error Variance in Stepwise Regression." *Journal of Economic and Social Measurement*, Vol. **31, Nos. 1-2**, 89–102.

Leinweber, David J. (2007). "Stupid Data Miner Tricks: Overfitting the S&P 500." *The Journal of Investing*, **16**(1), 15–22.

Tan, Pang-Ning, Michael Steinbach, and Vipin Kumar. (2006). *Introduction to Data Mining*. Boston: Addison-Wesley.

Vuk, Miha, and Tomaz Curk. (2006). "ROC Curve, Lift Chart and Calibration Plot." *Metodoloski Zvezki*, **3**(1), 89–108.

# Exercises

1. Create 30 columns of random numbers and use stepwise regression to fit them (along with S&P500) to the McDonalds return data.

   To create the 30 columns of random normal, first copy the McDonalds72 data set to a new file, say, McDonalds72-A. Open the new file and delete the 30 columns of "stock" data. Select **Cols→Add Multiple Columns**. Leave the Column prefix as Column and for How many columns to add? enter 30. Under Initial Data Values, select **Random**, then select **Random Normal**, and click **OK**.

   After running the stepwise procedure, take note of the RSquared and the number of "significant" variables added to the regression. Repeat this process 10 times. What are the highest and lowest $R^2$ that you observe? What are the highest and lowest number of statistically significant random variables added to the regression?

2. Use the churn data set and run a logistic regression with three independent variables of your choosing. Create Lift and ROC charts, as well as a confusion matrix. Now do the same again, this time with six independent variables of your choosing. Compare the two sets of charts and confusion matrices.

3. Use the six independent variables from the previous exercise and develop a neural network for the churn data. Compare this model to the logistic regression that was developed in that exercise.

4. Use the Freshmen1.jmp data set. Use logistic regression and classification trees to model the decision for a freshman to return for the sophomore year. Compare the two models using Lift and ROC charts, as well as confusion matrices.

5. Reproduce Table 10.4. Open a new Excel spreadsheet and copy the variables Return on McDonalds and Predicted Return on McDonalds into columns A and B, respectively. In Excel perform the following steps:

   a. Create the residuals in column C, as Return on McDonalds – Predicted Return on McDonalds.

   b. Create the squared residuals in column D, by squaring column C.

   c. Create the absolute residuals in column E, by taking the absolute value of column C.

   d. Calculate the in-sample MSE by summing the first 60 squared residuals (which will be cells 2-61 in column D). Then divide the sum by 60.

   e. Calculate the in-sample MAE by summing the first 60 absolute residuals (which will be cells 2-61 in column E). Then divide the sum by 60.

f.  Calculate the out-of-sample MSE by summing the last 12 squared residuals (cells 62-73 in column D). Then divide the sum by 12.

g.  Calculate the out-of-sample MAE by summing the last 12 absolute residuals (cells 62-73 in column E). Then divide the sum by 12.

h.  Calculate the in-sample correlation between Return on McDonalds and Predicted Return on McDonalds for the first 60 observations using the Excel CORREL( ) function.

i.  Calculate the out-of-sample correlation between Return on McDonalds and Predicted Return on McDonalds for the last 12 observations using the Excel CORREL( ) function.

---

[1] This example is based on Foster and Stine (2006). We thank them for providing the McDonalds and S&P 500 monthly returns.

# Chapter **11**

## Telling the Statistical Story

# From Multivariate Data to the Modeling Process

As you have read the early chapters, we hope you have come to realize we feel strongly that before discussing predictive analytics or performing a modeling project, one needs to understand how to deal with multivariate data. That is one of the book's main objectives. In particular, one needs a foundation beyond the univariate/bivariate analysis taught and learned in a basic statistics course to understand some of the issues when dealing with real-world, that is, multivariate, data. Hopefully, we have achieved that goal in the previous chapters. Now, we feel you are better prepared to understand data mining/predictive analytics/predictive modeling and to conduct a modeling project. The objective of this chapter is to provide a basic overview to data mining/predictive analytics/predictive modeling and to the modeling process using JMP.

For the past 25 years, the big buzzword in the BA area has been data mining. The roots of data mining techniques run deep and can be traced back to three areas—statistics, artificial intelligence (AI), and machine learning. All data mining tools and techniques have a strong foundation based on classical statistical analysis. In the 1970s and 1980s, AI techniques based on heuristics that attempted to simulate human thought processes were developed. Subsequently the field of machine learning, which is the union of statistics and AI, evolved. An example of machine learning is a computer program that learns more about the game of chess as it plays more and more games.

Two areas of early successful application of data mining have been credit card fraud detection and customer relationship management (CRM).

Based on analyzing customers' historical buying patterns, data mining models identify potential credit card fraud in transactions that are out of the "norm." For example, let's say you have never traveled to South America, but you happen to want to go to the World Cup in Brazil in 2014. So, you book a flight to Brazil and charge it on your credit card. While in Brazil you thought it would be nice to go on a few side trips—for example, to see Iguassu Falls. You then book a flight and accommodations and tour package with your credit card. Subsequently, you receive an e-mail from your credit card company saying that your card transactions are temporarily suspended and please contact them. The credit card company wants to make sure your card has not been stolen because you were making significant purchases outside your normal spending pattern. Data mining models are used to target such behavior.

The other area of early successful data mining applications is customer relationship management (CRM). CRM is a process/business strategy taken by companies to improve overall customer satisfaction, especially for their best customers. For example, large companies with multiple product offerings may have several customers that buy products across the company's product line. However, each division of the company may have separate sales and support staff as well as their own independent database. A CRM solution to this situation would be one company-wide database that allows everyone in the company to access the data on a particular customer, which improves customer satisfaction and promotes cross-selling opportunities.

SAS provides numerous customer stories of successful data mining/statistical applications that use JMP in several industry areas (aerospace, conservation, education, energy, genomics, government, health care, manufacturing, pharmaceuticals, and semiconductor) and by statistical application areas (see JMP Customer Stories).

# What Is Data Mining?

A broad definition of data mining is a process of finding patterns in data to help us make better decisions. Or more simply, as a good old friend of ours would say, it is mining data. In a nutshell, he is basically right. Furthermore, let's put forward this quote (Ewen, 1996) from the *New York Times*:

> Probably at no time in the last decade has the actual knowledge of consumer buying habits been as vital to successful and profitable retailing as it is today.

However, this statement was written in 1931. So data mining is not new, and successful decision makers have always done this. Then, why has the area of data mining grown so much recently? What has changed? The change has been the confluence of the three areas of data mining: (1) statistics, AI, and machine learning; (2) the exponential increase in our computer power; and (3) our scale of data accumulation have amplified this new area called data mining.

Nonetheless, data mining is not the current buzzword anymore. It has been replaced by the terms predictive analytics and predictive modeling. What is the difference in these terms? As we discussed, with the many and evolving definitions of business intelligence in Chapter 1, these terms seem to have many different yet quite similar definitions. One SAS expert defines these terms as follows:

> Data mining has been defined in a lot of ways, but at the heart of all of those definitions is a process for analyzing data that typically includes the following steps:
>
> □ Formulate the problem
> □ Accumulate data.
> □ Transform and select data.
> □ Train models
> □ Evaluate models.
> □ Deploy models.
> □ Monitor results.

Predictive analytics is an umbrella term that encompasses both data mining and predictive modeling–as well as a number of other analytical techniques. I define predictive analytics as a collection of statistics and data mining techniques that analyze data to make predictions about future events. Predictive modeling is one such technique that answers questions such as:

- Who's likely to respond to a campaign?
- How much do first-time purchasers usually spend?
- Which customers are likely to default?

Predictive analytics is a subset of analytics, which more broadly includes other areas of statistics like experimental design, time series forecasting, operations research and text analytics [1].

# A Framework for Predictive Analytics Techniques

Even with all these buzzword terminology variations, there appear to be two major characterizations of the terms data mining, predictive analytics, and predictive modeling. One point of view to view them as a collection of advanced statistical techniques. The other major point of view is to view them as a modeling process.

From the first point of view, several approaches have been used to classify data mining/predictive analytics/predictive modeling [2]. We categorize these predictive analytics techniques into supervised (directed) or unsupervised (undirected) learning techniques as shown in Figure 11.1.

With the unsupervised learning techniques, there is no target, or dependent variable(s). Beyond the discovery tools and multivariate techniques of Principal Component Analysis, Factor Analysis, and clustering that we discussed in this book, an example of an unsupervised learning predictive analytics technique is association rules (or market basket analysis or affinity grouping). With the association rules technique, we try to identify which things (in most cases, products) go together. For example, when you go grocery shopping, which products are sold together? An example would be milk and cereal or the unexpected classic data mining example of diapers and beer.

**Figure 11.1  A Framework for Predictive Analytics Techniques**

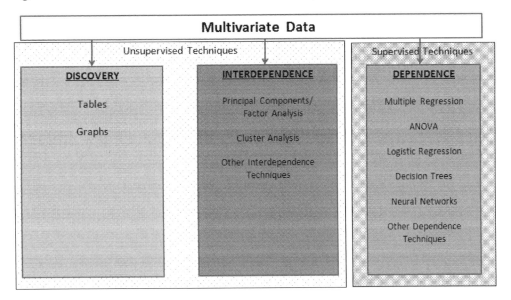

With supervised learning techniques, the goal is to develop a model that describes one (or possibly more than one but in most situations only one) variable of interest. The goal is to establish some relationship(s) among the variables. We have examined several such supervised techniques in this book: regression, logistic regression, ANOVA, decision trees, and neural networks. The decision tree and neural networks techniques are usually considered supervised learning predictive analytics techniques.

Finally, notice that in classifying and listing these predictive analytics techniques (Figure 11.1), we do include the basic statistical tools and techniques that you learned in the introduction to statistics as well as the multivariate techniques we discussed in this book. These tools and techniques are also part of predictive analytics and the modeling process.

# The Goal, Tasks, and Phases of Predictive Analytics

The goal of data mining/predictive analytics/predictive modeling/advanced statistical techniques, supervised or unsupervised, is to extract information from the data. The six main tasks of predictive analytics are listed below with their associated activities:

- discovery: describing, summarizing, and visualizing the data and developing a basic understanding of their relationships.
- classification: classifying each object to a predefined set of classes or groups.
- estimation: similar to classification but the dependent/target variable is continuous.
- clustering: segmenting each object into a number of subgroups or clusters. The difference between classification and clustering is that with clustering the classes or groups are not predefined but are developed by the technique.
- association: determining which items go together (*e.g.*, which items are brought together/concurrently).
- prediction: identifying variables that are related to (a) variable(s) so as to predict or estimate their future values.

The tasks of discovery, clustering, and association are all examples of unsupervised (undirected) learning. The other three tasks—classification, estimation, and prediction—are examples of supervised (directed) learning.

In this text, we have examined several of the fundamental data mining/multivariate techniques/advanced statistical techniques: discovery tools, clustering, Principal Component Analysis and Factor Analysis, ANOVA, regression, logistic regression, decision trees, neural networks, and model comparison. JMP is a comprehensive statistical and predictive analytics package. So, in addition to the JMP techniques/tools we discussed in the text, JMP also provides other predictive analytics/multivariate techniques such as conjoint analysis (in particular, discrete choice analysis) as well as several other statistical techniques.

So, what are the differences between statistics and data mining/predictive analytics/predictive modeling? This question is difficult to answer. First, both disciplines share numerous similar tools and techniques. However, both disciplines are much more than several tools and techniques. The major differences seem to lie in their objectives and processes. The broadening of the definition for predictive analytics from a collection of statistical techniques to a process is the second point of view[3] of predictive analytics.

Berry and Linoff (2004) define data mining as "a business process for exploring a large amount of data to discover meaningful patterns and rules". The phases of the data mining process are listed in Table 11.1. This process is not necessarily linear; that is, you do not always proceed from one phase to the next listed phase. Many times, if not most of the time, depending on the phase's results, the data mining project may require you to go back one or more phases. The process is usually iterative.

### Table 11.1  The Phases of the Data Mining Process and the Percentage of Time Spent on Each Phase

| Data Mining Process | |
|---|---|
| Project definition | (5%) |
| Data collection | (20%) |
| Data preparation | (30%) |
| Data understanding | (20%) |
| Model development and evaluation | (20%) |
| Implementation | (5%) |

As you can see from Table 11.1, what we have discussed in this book concerns only 20% of the time spent on a data mining project: model development and evaluation. While "data understanding" does require some use of statistics (scatterplots, univariate summary statistics, etc.), easily 50% of the analyst's time will be spent on the mundane and tedious tasks of data preparation and data understanding. These critical tasks are beyond the scope of this book. But we wish to note that very little is written about them, which makes learning about these topics difficult. There is a notable exception, though—the excellent book by Pyle (1999). We recommend it to anyone who wishes to understand the basics of data collection and understanding.

Often in statistical studies, the study's objectives are well defined, so the project is well focused and directed. The data is collected to answer the study's specific questions. A major focus of most statistical studies/processes is to draw inferences about the population based on the sample.

While on the other hand in many predictive analytics projects, besides having a significantly large data set, in many cases, the data is the entire population, thus, making statistical inference a moot point. The data in a predictive analytics project is rarely collected with a well-defined objective of analysis, and it is usually retrieved from several data sources. As a result, unlike most statistical studies, the data must be integrated from these different sources and appropriately aggregated. Just like statistical studies, the data in a data mining project must be cleaned and prepared for analysis. However, due to the numerous sources of data and the usually larger number of variables, this phase of the process is much more labor intensive. Both processes share the same concern: to develop an understanding, description, and summary of the data.

The primary phase of the data mining/predictive analytics modeling process, which many people would define as data mining/predictive analytics (the first point of view), is the model development and evaluation phase. This phase may account for only about 20% of the project's overall efforts (mainly because of the large amount of effort to integrate and prepare the data).

SAS Institute Inc. developed a systematic approach to this phase of the data mining process called SEMMA (Azevedo and Santos, 2008):

**S—Sample:** If possible (that is, if you have a large enough data set), extract a sample that contains the significant information yet is small enough to process quickly. The part of the data set that remains may be used to validate and test the model developed.

**E—Explore:** Use discovery tools and various data reduction tools to further understand data and search for hidden trends and relationships.

**M—Modify:** Create, transform, and group variables to enhance the analysis.

**M—Model:** Choose and apply one or more appropriate data mining techniques.

**A—Assess:** Build several models using multiple techniques; evaluate, assess the usefulness, and compare the models results. If a small portion of the large data set was set aside during the sample stage, validate and test the model.

Once the "best" model is identified, the model is deployed, and the ROI from the data mining process is realized.

The objective of the model development and evaluation phase is to uncover unsuspected but valuable relationships. So you search until you find a model that fits the data set arbitrarily well, so that it is not overly complex and the model does not overfit the data. Statisticians become concerned with such a data-driven analysis approach to obtain a good fit because they are aware that such a search could lead to relationships that happen purely by chance. Unlike most statistical studies, predictive analytics projects are less focused on statistical significance and more on the practical importance—i.e., on obtaining answers that will improve decision making. Nevertheless, even though objectives and processes may differ, the bottom line of statistical studies and data mining projects is to learn from the data.

We hope this book has provided you with a foundation to conduct a statistical study (or a predictive analytics project) and planted the seeds on how to write a statistical story. Happy story-telling!

# References

Azevedo, A., and M. F. Santos. (2008). "KDD, SEMMA and CRISP-DM: a parallel overview," *Proceedings of the IADIS European Conference on Data Mining.* 182–185.

Berry, M. J. A., and G. S. Linoff. (2004). *"Data Mining Techniques: For Marketing, Sales, and Customer Relationship Management."* 2nd Ed. Indianapolis: Wiley Publishing, Inc. 2.

Ewen, Stuart. (1996). *PR!: A Social History of Spin.* New York, NY: Basic Books. 184.

JMP Customer Stories. http://www.jmp.com/software/success/alphabetical.shtml.

Pyle, Dorian. (1999). *Data Preparation for Data Mining.* San Francisco: Morgan Kaufmann.

---

[1] http://www.sas.com/news/sascom/2010q1/column_tech.html
[2] Remember that we use these terms interchangeably, to mean the same thing.
[3] This broadening definition of predictive analytics is reflective of the maturity of the discipline.

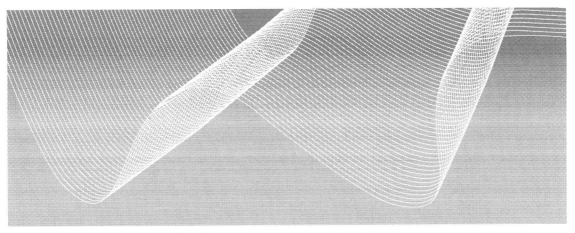

# A p p e n d i x

## Data Sets

At the end of most chapters, we have several exercises usually using (a) data set(s) used in the chapter or perhaps in another chapter. The main purpose of these exercises is to improve and/or expand on the mechanics of the chapter techniques. On the other hand, in doing so, we believe we are not improving the most difficult aspect of the statistical problem-solving process—deciding whether a technique is appropriate or not. We have taken a twofold approach to address this critical problem-solving step.

First, we have eight smaller data sets that can be assigned at the end of Chapters 2–9. Second, we also provide six rich case data sets—either with numerous observations and/or with numerous variables. These data sets, in general, would require more time to analyze and could be appropriate for semester-long assignments. In either case, both types of data sets are from real-world data. In some cases, the data set is not appropriate for the technique(s) covered in the chapter. For other situations, it is very appropriate, and the results may or may not provide any real benefit. We leave it up to the instructor as to how to assign these data sets.

In the next section, we provide a brief description of each data set.

# Smaller Data Sets

### City Ranking
### File: CityRanking.jmp

We see it all the time—a listing of the top ten cities to live in. How do they come up with these rankings? We have 367 metropolitan areas and several descriptive variables: **Population**, **Cost of Living**, **% Creative Class**, Median household income (**Med HH Income**), percentage income growth (**% Inc Growth**), **State**, and metropolitan area (**Metro Area**). How do these variables impact a city becoming ideal? And how do we rank the cities, from best to worse?

### Credit Card Statistics
### File: Credit Card Stats.jmp

We have 423,000 observations from a credit card company on their customers' purchasing patterns by month for the year 2009. The variables include:

> **accountnumber**, **Year_Mo**, **sum**, **productname**, **segmentdescription**, **categoryid**, and **merchantname**. Can you find any patterns or relationships that affect credit card sales?

### Crime Data
### File: Crimedata.jmp

The FBI compiles an annual report of the volume and rate of crime offenses for the nation, the states, and individual agencies. This report also includes arrest, clearance, and law enforcement employee data. The entire list of variables is shown in the Table below.

| Region | | Murder Rate | | Total (Violent+Property) | | Burglary |
|---|---|---|---|---|---|---|
| State | | Rape Rate | | Violent | | Larceny |
| Year | | Robbery Rate | | Property | | MVTheft |
| Population | | Agg-Aslt Rate | | Murder | | NewTotal |
| Total Rate | | Burglary Rate | | Rape | | New Violent |
| Violent Rate | | Larceny Rate | | Robbery | | New Property |
| Property Rate | | MVTheft Rate | | Agg_aslt | | |

The data set has this annual data from 1973 to 1999. Can you find any patterns or relationships in this crime data that will be helpful to the FBI and/or local police?

### Retention

### File: Freshman

A major issue that most universities face is minimizing the number of students who drop out, leave, or transfer. The data set contains information on 100 college students that have just completed their freshmen year. Variables in the file include College **GPA**, **Miles from Home**, **College** (within the university), **Accommodations** (dormitory or off-campus housing), **Years Off** (time off between high school and college), **Part-time Work Hours**, **Attends Office Hours**, and **High School GPA**. The university hopes to understand which variables contribute to whether a student will fail during freshman year and leave the school or succeed and return for their sophomore year.

### Health Care Trends

### File: Healthtrends

Trends in health care are extremely useful to public policy agencies and health care industry companies such as pharmaceutical companies. Several companies are repositories of the data from different aspects—e.g., physician visits or prescriptions—of the health care industry. The data set contains 3129 records, which are at a snapshot macro level data set of physician visits for three months. The variables included are medical procedure (**Procedures**), medical diagnosis (**Diagnosis**), and patient count by week. Given this data set, answer the following questions:

1. Why patients are being treated—diagnoses

2. How are patients being treated—procedures

3. Are there trends in treatment

### Massachusetts Housing

### File: MassHousing

Although lagging behind business, federal and state governments are increasing their use of business analytics. In this data set, we have, for 506 towns in the metropolitan Boston area, the crime rates and the following associated variables:

**crim**: per capita crime rate by town

**zn**: proportion of residential land zoned for lots over 25,000 sq. ft.

**indus**: proportion of non-retail business acres per town

**chas**: Charles River dummy variable (1 if tract bounds river; 0 otherwise)

**nox**: nitric oxide concentration (parts per 10 million)

**rooms**: average number of rooms per dwelling

**age**: proportion of owner-occupied units built prior to 1940

**distance**: weighted distances to five Boston employment centers

**radial**: index of accessibility to radial highways

**tax**: full-value, property-tax rate per $10,000

**pt**: pupil-teacher ratio by town

**b**: 1000(Bk – 0.63)2, where Bk is the proportion of blacks by town

**lstat**: % lower status of the population

**mvalue**: median value of owner-occupied homes is $1000

The state of Massachusetts is interested in understanding which factors may have an impact on crime rates.

### Equality Promotion

**File: Promotion**

To be eligible for promotion to lieutenant and captain, according to a fire department's union-weighted score, 60% of the candidates are assigned a written exam and 40% an oral. An overall score of 70% must be achieved. The data set has the results for all 118 firefighters that took the exam: 77 for promotion to lieutenant, and the remaining 41 for captain. The variables in the file are:

**Race**: W = white, H=Hispanic, B=black

**Position**: Captain or lieutenant

**Oral**: Oral exam score

**Written**: Written exam score

**Combine**: Weighted total score, with 60% written and 40% oral

At the time of the exam, 8 lieutenant and 7 captain positions were available. First, assist the fire department management and identify the top 10 candidates for each position. Second, the exams were not certified, so there may be concern about fairness and, in particular, concern about reverse discrimination.

### Titanic Survivors

**File: Titanic Passengers**

What may have affected the survival of passengers of the Titanic? The data set contains information on 1309 individual passengers (not the crew, only passengers). The passenger variables are:

> **Passenger Class, Survived, Name, Sex, Age, Siblings and Spouses, Parents and Children, Ticket #, Fare, Cabin, Port, Lifeboat, Body, Home/Destination, and Midpoint age.**

The question to be addressed is: Did any of the passenger variables have an effect on their survival?

# Large Case Data Sets

### Apples

### File: Applessurvey.jmp

Over the past several years, the fresh fruit and vegetable industries have experienced significant increase in the propensity of consumers to purchase local and organic food. In particular, from 1997 to 2008, the consumption of organic foods and beverages increased from $3.6 billion to $21.1 billion. Several interrelated factors have been the driving forces behind these trends. Examples are: concern for healthy foods, desire for better tasting foods, concern over chemicals/pesticides in food, and simply providing support to local industry. One of the most frequently produced locally and purchased fruits are apples. So, the study focused on apples.

An online survey of adult residents of Pennsylvania was conducted in 2009 and resulted in 1224 completed surveys. Due to Pennsylvania's diversity of urban, suburban, and rural environments, industrial and agricultural commerce, and additionally, since the state is a major producer and consumer of apples, the state is viewed as a good representative sample. The major objectives of the survey were to evaluate the market opportunities and profitability of organic farming and to identify the factors that influence consumer purchasing of organic apples. In the file **Applesurveya.xlsx**, are two worksheets, one with the survey results (survey responses) and one that describes the survey questions (Survey questions). The survey results are also in **Applessurvey.jmp**. In the Word file **Organic Apple Surveya**, is the survey instrument.

**Note:** There is no question 53; Also, 1—Yes and 2—No; and 1—checked and 0—not checked.

### Bank Churn

### File: CC Churn.jmp

The highly competitive bank industry has been a leader in using business analytics. Two areas of application have been to use their data to identify characteristics in retaining their current customers and in obtaining new customers. Davenport and Harris noted in their book *Competing on Analytics*[1], that Capital One is one of industries' leaders in using their data. Common measures used in the industry are churn or churn rate. Churn is either yes or no: yes if you leave the current service and no if you stay. Churn rate is the percentage of customers that leave/stop using a service during a certain period of time. In the JMP file **CC Churn.jmp**[2] is a bank's customer data with 245,465 observations, which contain several descriptive characteristics: **Churn Flag**, **cust id**, **Average Daily Balance**, **Interest Paid**, **Cash Advances**, **Balance Transferred**, **Marital Status**, **Occupation Group**, **Age of Account (Months)**, **Age Group**, **LTV Group** (life time value group), **Bill Cycle**, **Customer Type**, **Gender**, **Customer Value**, and **Credit Limit**.

Can this information be used to predict churn?

### Enrollment

### File: Enrollment.jmp

University admissions offices have dramatically changed over recent years in how they contact, communicate, and uncover perspective students. For most students, their initial contacts with the university are over the Internet. Let it be by the university's website, Facebook, or even Second Gen. The old days of just visiting certain high schools and going to college fairs is simply just not enough. Every admissions office is exploring ways to obtain a competitive advantage. The data set **Enrollment.jmp** has 39,441 observations for Fall 2006 to Fall 2010. This undergraduate admissions data is from a small (total student population, undergraduate and graduate, of just less than 9000) university. The variables are:

| | | Merit-Based Financial Aid |
|---|---|---|
| Academic Period | Gender Description | |
| Unique ID | Secondary School | Residency Indicator |
| State Province | High School GPA | Common Application-Paper |
| Nation Description | Act English | Common Application |
| Student Level | Act Math | Saint Joseph's Online Application |
| Student Population | Act Reading | Common Application Upload |
| Application Date | Act Science Reasoning | Saint Joseph's Paper Application |
| Admissions Population Description | Act Composite | Pre-Dental |
| Residency Description | Sat Verbal | Pre-Law |
| College Description | Sat Mathematics | Pre-Med |
| Major Description | Sat Total Score | Pre-Veterinarian |
| Applied | ACRK Index | |
| Admitted | Institutional Aid Offered | |

*(continued)*

| Enrolled | Class Rank | | |
|---|---|---|---|
| Legacy Description | Class Size | | |
| Citizenship Description | Class Rank Percentile | | |
| Religion Description | Nation Of Birth Description | | |
| SOC | Admissions Athlete | | |
| Ethnicity | Need-Based Financial Aid | | |

More university admissions offices are analyzing their admissions data to better understand not only the students that do enroll at the university, but, also, those that do not enroll. Can you help them?

**Home Equity**

**File: hmeq.jmp**

The banking industry is one of the leaders in applying business analytics as part of their operation. The data available is from 5960 customers in the file **hmeq.jmp**[3]. The variables are:

**Default**: 1 = defaulted on loan; 0 = paid load in full

**Loan**: Amount of loan requested

**Mortgage**: Amount due on existing mortgage

**Value**: Current value of property

**Reason**: Reason for the loan request—HomeImp = home improvement; DebtCon = debt consolidation

**Job**: Six occupational categories

**YOJ**: Years at present job

**Derogatories**: Number of major derogatory reports

**Delinquencies**: Number of delinquent credit lines

**CLAge**: Age of oldest credit line in months

**Inquiries**: Number of recent credit inquiries

**CLNo**: Number of existing credit lines

**DEBTINC**: Debt-to-income ratio

In this scenario, a bank would like to use their customer information to assist them in determining whether or not they should approve a home equity loan.

**Pharmaceuticals**

**File: Pharm.xls**

EMD Research is one of the leading healthcare market research companies providing pharmaceutical companies and investors with forecasts and trends of drug usage. In the file **Pharm.xls** are weekly data for five cholesterol drugs (Crestor, Lipitor, Vytorin, Zetia, and Zocor) and for several strengths (i.e., dosage levels) for 2 ¼ years[4]. Besides trying to possibly improve their 3-month forecasts, can you discover any significant trends or findings in the cholesterol drug market?

**Cell Phone Churn**

**File: churn.jmp**

Cell phone companies are constantly bombarding us with new phones, programs, and other offerings so that we would leave our current company and change to their company. A term describing such behavior is churn. In the file **churn.jmp**, we have a data set from a cell phone company with data describing 3,333 customers and their cell phone characteristics[5]:

| | | | |
|---|---|---|---|
| **Account Length** | | **Eve Mins** | |
| **Area Code** | | **Eve Calls** | |
| **Phone** | | **Eve Charge** | |
| **Int'l Plan** | | **Night Mins** | |
| **VMail Plan** | | **Night Calls** | |
| **E_VMAIL_PLAN** | | **Night Charge** | |
| **D_VMAIL_PLAN** | | **Intl Mins** | |
| **VMail Message** | | **Intl Calls** | |
| **Day Mins** | | **Intl Charge** | |
| **Day Calls** | | **CustServ Calls** | |
| **Day Charge** | | **Churn?** | |

Using this data, can you assist this cell phone company in predicting churn?

---

[1] *Competing for Analytics: The Science of Winning*, Thomas H. Davenport and Jeanne G. Harris, Harvard Business School Press, 2007.

[2] Thanks to Chuck Pirrello of SAS for providing the data set.

[3] Thanks to Tom Bohannon of SAS for providing the data set.

[4] *Competing for Analytics: The Science of Winning*, Thomas H. Davenport and Jeanne G. Harris, Harvard Business School Press, 2007.

[5] Thanks to Tom Bohannon of SAS for providing the data set.

# Index

## A

absolute error measures  226, 227, 244

absolute penalty fit method  211

accuracy characteristic of prediction model  231

activation function, neural network  203–204

Adjusted $R^2$  75, 76

affinity grouping  252

agglomerative algorithm, clustering  154–163

AI (artificial intelligence)  250

AIC/AICc (Akaike information criterion) approach
          76, 184, 195–196

Analyze command
        Fit Line  65
        Fit Model  67–69, 95
        Fit Stepwise  74
        Fit Y by X  31–33, 65, 83–84

ANOVA (analysis of variance)
        one-way/one-factor  83–96
        process  82–83
        two-way/two-factor  97–102

area under the curve (AUC)  235, 236, 237

artificial intelligence (AI)  250

association rules  252

association task, predictive analytics  254

AUC (area under the curve)  235, 236, 237

average linkage method, distance between clusters
          154–155

Axis Titles, Histogram Data Analysis  23

## B

BA (business analytics)  3–5, 250

bar chart  59–61

Bayesian information criterion (BIC)  76

bell-shaped distribution  18

BI (business intelligence)  3–5

bias term, neural network  203

BII (business information intelligence)  3–5

binary dependent variable  104, 221–222, 230–237

binary vs. multiway splits, decision tree  181

bivariate analysis  6, 31–36, 124–133

BMI (business modeling intelligence)  3–5

boosting option, neural network predictability  210,
          216–217

BSI (business statistical intelligence)  3–5

bubble plot  53–55

business analytics (BA)  3–5, 250

business information intelligence (BII)  3–5

business intelligence (BI)  3–5

business modeling intelligence (BMI)  3–5

business statistics intelligence (BSI)  3–5

## C

categorical variables
        *See also* ANOVA
        deciding on statistical technique  26, 28–29
        decision tree  180, 181–192
        graphs  45–46, 50–51, 52
        neural network  208, 212–213
        regression  76–82
        tables  42

causality  39

central limit theorem (CLT)  18–24

centroid method, distance between clusters  154–155,
          164–166

chaining in single linkage method, distance between
          clusters  154

chi-square test of independence  185

churn analysis  122–133

classification task, predictive analytics  254

classification tree  181–192, 237, 239–240

cleaning data for practical study  8

CLT (central limit theorem)  18–24

cluster analysis
        credit card user example  152–153
        definition  152
        hierarchical clustering  154–163, 177
        *k*-means clustering  154, 164–177
        regression, using clusters in  164

Cluster command  156–157, 159

Clustering History, Cluster command  159

clustering task, predictive analytics  254

coefficient of determination (RSquare or $R^2$)  66

Color Clusters, Cluster command  157

Column Contributions, decision tree  198

complete linkage method, distance between clusters
          154–155

confusion matrix
        binary dependent variable model comparison
          230–231
        bivariate analysis contingency table  35

# ACCELERATE YOUR SAS® KNOWLEDGE WITH SAS BOOKS

Learn about our authors and their books, download free chapters, access example code and data, and more at **support.sas.com/authors**.

Browse our full catalog to find additional books that are just right for you at **support.sas.com/bookstore**.

Subscribe to our monthly e-newsletter to get the latest on new books, documentation, and tips—delivered to you—at **support.sas.com/sbr**.

Browse and search free SAS documentation sorted by release and by product at **support.sas.com/documentation**.

Email us: sasbook@sas.com
Call: 800-727-3228

**THE POWER TO KNOW®**

Made in the USA
Lexington, KY
17 May 2013